SEE THE WORLD THROUGH

PATTERNS

SEE THE WORLD THROUGH
PATTERNS

Robert Barkman, PhD

Clear Message Press
Clear Message Media
Woodbridge, VA 22191
www.clearmessagemedia.org
@CMMLLC

Printed in the United States of America
First Paperback Edition: April 2022

Published by Clear Message Press, an imprint of Clear Message Media, LLC. The Clear Message Press name and logo is a trademark of Clear Message Media.

ISBN: 978-0-9981480-1-4 (paperback)
eISBN: 978-0-9981480-2-1 (e-book)

Library of Congress Control Number: 2022934644

This book is dedicated to
Richard "Dick" Konicek-Moran (1931-2019)

Who inspired me and the hundreds of other teachers like me to give students ownership in their own learning by finding out what students already know and build upon it.

CONTENTS

CHAPTER 6

PREFACE AND PURPOSE

Scientific advancement is the discovery of new patterns. We expect meaning in the patterns we see because, in a random universe, it takes energy to create order. So, when we see a particular pattern, we expect that through investigation we can identify the force that caused it. That's how we learn new things.

Greg Satell, The Science of Patterns, *Forbes*, May 1, 2015

Readers of this book learn from a variety of stories and exercises designed to engage readers with what patterns are, how diverse they are, and what we can discover from them. Because we all are pattern smart in different ways, the book describes how readers can recognize this skill in themselves and learn to see patterns that may have gone unnoticed with the unaided eye.

Why Is Pattern Recognition Important? What Is the Educational Significance?

Learning about pattern recognition is for people who may not know they need it or realize they have it. Pattern recognition is what gave us the evolutionary edge over animals and helped to make us human. Advancements of our civilization are often credited to the discovery of new patterns. We often operate on autopilot, unaware that our brains are constantly searching for patterns and deciding how to respond to them. Readers will be surprised to learn that seeking out patterns is probably the most important thing that the brain does. Few realize that humans can recognize many types of patterns almost as fast as any pattern recognition machine, and transform these into actions. How we refine, shape, and improve our pattern-recognition skills is key to how much longer we will have the evolutionary edge over machines.

Patterns are powerful. They set up expectations, make connections, and inspire burning questions. They can be events that regularly repeat themselves, trends in which events rise or fall over a prolonged period, relationships that create new connections, or they can emerge from seeing the larger picture. They can be outliers, events that fall outside the norm, or newly defined patterns called fractals. Together, pattern recognition inspires both scientists and engineers, which can lead to new discoveries, breakthrough ideas, and innovative concepts.

Seeking patterns changes the way we see, think, and succeed by becoming alert to our brains' abilities to recognize patterns. It challenges the idea that discovery is accidental. Far more common is that discovery is inspired by a pattern that piqued the curiosity of someone; often that pattern was overlooked by others. When we see a pattern, we want to explain why it occurs that way. It advances the idea that genius is seeing patterns where others see only chaos. Because we are all pattern smart in different ways, the book guides readers to self-discover how they are pattern smart.

What Significance Does Pattern Recognition Have to Science, Technology, Engineering, and Math (STEM)?

My idea for the book is that pattern recognition is a skill that has value beyond the sciences. I was thinking about the teachers who teach interdisciplinary science, which links math, language, and even art to science. The well-known STEM studies, for example, integrates technology, engineering, and math with science. Pattern recognition plays a role in each of these. Patterns are the hooks on which to hang the big ideas common to all the STEM fields: science, technology, engineering, and math.

The point of departure between science and everyday observation is the collection of data and close attention to patterns (National Research Council, 2005). Recognizing patterns, for example, is the first step that inspires questions about how and why the pattern occurred. Scientists seek to explain observed patterns and look for the similarity and diversity within them. To put order in chaos, scientists start to organize objects into categories. Patterns emerge upon classifying the objects. The goal for students is to recognize, classify, and evaluate patterns. Start by looking at data; for example, record seasons, temperature, or phases of the Moon. Engineers look for and analyze patterns, too. For example, they may diagnose patterns of failure of a designed system under test in order to improve the design. The ways in which data are represented can facilitate pattern recognition. This practice can lead to the development of a mathematical representation, which can then be used as a tool to seek an explanation for what caused the pattern to occur (National Research Council, 2012).

A lot of human scientific and technological progress over the span of recorded history has been related to discerning patterns. People early in our history noticed that the Sun and Moon both had regular periodicity to their movements, leading to models that ultimately changed our view of our place in the universe (National Research Council, 2012). More recently, patterns recognition continues to inspire new discoveries, new ways of doing things, and new inventions.

Learn how because of Dr. Allen Steere's attention to seasonal and geographical patterns, Lyme disease was discovered. An interview with Robert Ballard reveals how his attention to patterns predicted where the Titanic would be found. Discover how Clarence Birdseye created an entire food industry by observing the pattern of behavior of fish pulled through the ice by the Inuit Indians of Labrador. Computer engineer and cofounder of Intel, the late Gordon Moore, noticed that the number of transistors per square inch on integrated circuits doubled every year since their invention. Moore predicted that this trend, now coined Moore's Law, would continue into the foreseeable future. In my interview with him, even he was surprised that Moore's Law, now over 50 years old, continues today.

Connecting to *A Framework for K–12 Science Education* and the *Next Generation Science Standards*?

See the World Through Patterns supports the crosscutting concepts now elevated to the importance of core ideas and science and engineering practices. The first of the seven concepts are patterns. Why patterns? When you see them, they can be life changing. They can even make you smarter. Recognizing a pattern is like looking through a telescope for the first time. You see things through new eyes that you have never seen before. When one of my students looked through a telescope at Saturn for the first time, his first words were "d... amazing." The planet's famous rings inspire many first-time astronomers, and that can fuel an enthusiasm and a lifelong interest in astronomy.

Although crosscutting concepts are fundamental to an understanding of science and engineering, "students have often been expected to build such knowledge without any explicit instructional support" (A Framework for K–12 Science Education). Moreover, the research base on learning and teaching the crosscutting concepts is limited. The book See the World Through Patterns helps to fill this need by offering teachers and their students as well as all interested readers a resource that prepares them to understand the role that pattern recognition played in past discoveries and innovations and its significance to readers' home and workplace.

Readers will find firsthand information learned through interviews from leaders in their respective fields who share their insights into the discovery process and the role pattern recognition played in their own breakthroughs. The book, where appropriate, makes connections to the Next Generation Science Standards and gives exercises that offer hands-on learning about the significance of patterns to scientific discovery.

How Does the Book Relate Patterns to the Brain and Learning?

Part of what makes us human is the need for our brains to search for meaning. Good educators understand this fact and create learning opportunities that honor this reality.

Students who are given complex patterns to discover and interpret send their brains on this important search for meaning. At the same time, students are building the connections so vital to higher learning and critical-thinking skills.

A Patterned Existence

Patterns are observations organized into meaningful categories by the observer. When students seek patterns in the world around them, they see order instead of chaos, which builds confidence in their understanding of how the world works and gives them a feeling of control.

Charles Darwin drew upon his understanding of patterns when he synthesized his evolutionary theory from observations of life pieced together from his voyage. Our laws of heredity are products of Gregor Mendel's careful recording of the patterns of inheritance of pea plant traits. Rachel Carson's observation of the patterns of bird die-offs led her to identify the misuse of chemical pesticides at fault.

There are patterns to be discovered in numbers, people, musical scores, and even one's thoughts. Mathematics is often described as the language of patterns. Patterns in numbers made possible the discovery of pi and the invention of Pascal's triangle. The identification of Fibonacci numbers (1, 2, 3, 5, 8, 13, and so on) has led to the recognition of the presence of this serial pattern in pineapples, sunflowers, and pinecones. Through computer analysis, patterns known as fractals have been discovered in music. These patterns reveal that only a few musical notes are responsible for the basic melody of some of the world's greatest works. Thought patterns become recognizable when a person focuses on thoughts and emotions that repeat themselves. Therapists use this technique to improve a patient's control over stress and circular thinking.

Patterns can also be found in languages. Linguistic researchers have long been fascinated by the similarities of the German and Dutch version for the English word *fist: faust* and *vuist*. Common features in language such as these have been used to reconstruct the evolution of languages and the history of peoples.

Even when none are intended, patterns emerge. The ancients looked up at the stars, made spatial connections among them and created the familiar nighttime constellations. On a more earthly plane, we know baseball as a game of hits, runs, catches, and pitches. Seasoned players have learned to read movements as patterns that show when best to hit away, steal second, or pitch a slider.

Implications for Teaching

In view of how the brain learns, students should be given opportunities to discover patterns rather than memorize theories. By studying the structures of squirrel nests, the "Morse code" of fireflies, the patterns of energy and life on north- and south-facing slopes, students can rediscover important scientific concepts of adaptation, diversity, and energy for themselves. This active seeking fosters the need to know as well as persistence, respect for evidence, and sense of stewardship and care—which all characterize good science. After all, one way to define science is as the human attempt to account for patterns in nature. Through planned and patient observations, anyone can learn the skill of recognizing patterns.

The 5Es—Engage, Explore, Explain, Extend, and Evaluate

Because we all are gifted with the brain's innate ability to recognize patterns, the book describes how readers can recognize this skill in themselves and learn to see patterns that may have gone unnoticed with the unaided eye.

Follow these steps using a well-known model of inquiry, the BSCS 5E Instructional Model (5Es)—Engage, Explore, Explain, Extend, and Evaluate—to solve the following pattern puzzles.

- Engage and challenge readers to explore different patterns,

- Explore a concept,

- Explain and interpret the pattern,

- Extend the learning and Evaluate its significance by applying the concept to new situations (Biological Science Curriculum Study, n.d.). Because it builds on how science inquiry is practiced, it is an ideal model to explore patterns from people-smart ones to music-smart ones, for example, in an orderly way.

While the learning cycle captures how patterns can inspire discovery, formal training is not a prerequisite to use it. Part of what makes us human is the need for our brains to search for meaning. The brain is designed to perceive and derive meaning from patterns, and it resists having meaningless information imposed on it. *Pattern Seekers* understands this fact and tries to create learning opportunities that honor this reality. People who are given complex patterns to discover and interpret send their brains on this important search for meaning. At the same time, pattern seeking builds the connections so vital to higher learning and critical-thinking skills. When one seeks patterns in the world around them, they see order instead of chaos, which builds confidence in their understanding of how the world works and gives them a feeling of control. Curiosity, being a good observer, and persistence, even a passion to explain the pattern, however, are required. When these ingredients are mixed in the right proportions, discovery is in reach of everyone!

Patterns Are Puzzles Waiting to Be Solved

Whether they are concentric rings found in fish ears or historical stock prices that track repeating shapes, patterns are puzzles waiting to be solved. We wonder why they exist. They have meaning and purpose, but the meaning and purpose are not always recognized. Patterns with hidden meaning and answers waiting to be discovered can be found in any field of human endeavor: math, business, science, forensics, health, music...you name it. At the end of each chapter are a few patterns to pique your or your students' interests and inspire future inquiry.

ACKNOWLEDGMENTS

Having an idea and turning it into a book is as hard as it sounds. The experience is both internally challenging and rewarding. I especially want to thank the individuals that helped make this happen.

To Dawn, my wife, who is my best critic and cheerleader and who endured the many nights and weekends that my work took to do over the past 8 years. To her, I say thank you!

To the rest of my family, Cara, Taryn, Chris, Jeff, Cade, Brock, Bryson, Tate, and Lola who always brightened my day.

To my doctor, Dr. Bruce Johnson, and the caregivers of Dana Farber. Thanks to them, I have been able to continue my daily routines.

Thanks to Dr. Darren O'Neill. As my primary doctor, he was always responsive and attentive to my health care needs.

Thanks to Carole Hayward, publisher of Clear Message Press who took a chance with me. Her comments stretched my thinking and inspired me to reach higher. Her excellent editing and attention to detail produced a book we can all be proud of.

Special thanks to Rachel Ledbetter, the ever-patient former Managing Editor of NSTA Press. I greatly appreciate the advice and constructive criticism Rachel offered during the last roundup of the book. Thanks to Claire Reinburg, former Director of NSTA Press, for her feedback and encouragement and Will Thomas, Director of Art at NSTA, for his help to create the many images for this book.

Thanks to Dr. David Miller for suggesting patterns to research in medicine and Dr. Bill Harbison for putting me in touch with Dr. Allen Steere to interview.

Thanks to these people for reading my manuscript and offering feedback and encouragement when I began first writing the book. Reverend Gary DeLong, Steve Gelling, John Glenn, Paul Katz, Dr. Linda Marston, Linda Roger, Dr. Mary Allen, Dr. Jennifer Stratton, Pete Hitler, Tony Rodolakis, and George Keady.

I am grateful for the opportunity to have interviewed the following patterns seekers and value the time they spent sharing stories about their groundbreaking discoveries with me.

The impact of Moore's Law is everywhere, embedded in devices millions of people use every day. Even the late Dr. Gordon Moore was amazed that Moore's Law persists today.

Credited for identifying and naming Lyme Disease, Dr. Steere's keen detective work uncovered several patterns of infection that pointed to an insect cause of Lyme disease.

If you ever thought that Robert Ballard discovered the Titanic by accident, you will think differently after reading about his attention to patterns.

The question, "Who would ever build a stone wall in the middle of a forest," was answered by Dr. Schlobaum's close attention to landscape patterns.

Few kids enjoyed playing with numbers and looking for patterns, but Dr. Arthur Benjamin did. His skill of doing fast mental math earned him the title of "America's Best Math Whiz."

Smith College Professor James Middlebrook who taught architecture thought a lot about nature and experimented with plant and animal patterns to inspire his award-winning designs.

If asked whether he would like to play video games, play soccer, or solve a complex math problem, the world's top data scientist, Gilberto Titericz, would answer, "tell me about that problem." Gilberto's attention to patterns plays a big part in tackling large data problems.

Lost hikers, according to tracking sleuth Judy Moore, usually make the same three mistakes. Number 1, they make a wrong turn. Number 2, instead of backtracking, they take another wrong turn. Everything they do after that will get them further lost (number 3).

There are only a few people in the world known to be natural at detecting facial patterns that suggest someone is lying. Called the super lie detector, Renee Ellory, is one of them.

Reading the landscape and reconstructing the past takes certain skills and talents. According to Tom Wessels, they include the ability to see changes in large-scale patterns, develop hypotheses to explain the changes, and then look for small-scale evidence to support or reject the hypotheses.

Claire D'Amour Daley, Vice President of Massachusetts Big Y Supermarkets, says their company studies patterns of customer buying habits much like a scientist would. Their store layouts reflect these patterns.

Thanks also to Dave Bend, Opower Corporation; teacher Judy DeLong; mathematician Dr. Andrew Perry; and Dr. Judy Willis, neurologist and teacher; Springfield College librarian, Lynn Martin for enlightening me about subjects that they are considered expert.

To Page Keeley, Ron St. Amand, and Dick Konicek, who reviewed my proposal, and the four anonymous reviewers who critiqued my manuscript, I greatly appreciate your time and effort. The manuscript is in a much better place because of you.

INTRODUCTION

"There's something," Stu Harris said, pointing to the image on the screen. Before that moment, all they had seen were endless miles of mud. There on the screen were images of something human-made. "Bingo!" shouted one crew member. It was debris from the *Titanic* (Ballard, *Exploring the Titanic*, 1988). Robert Ballard's immediate reaction was, "cool, this is amazing," jumping up and down with his crew like children. Then, someone said, "You know, she sinks in 20 minutes." It was 2:00 in the morning, and the *Titanic* sank at 2:20. We all went, "Ah." That comment made us think, "Oh, wow, what are we doing celebrating?" The mood changed like someone hit a wall switch. It went somber. We went out of the room and had a private ceremony, then came back and never left that somber state again (Ballard, *The Discovery of the Titanic*, 2016).

Robert Ballard claimed in my interview with him that the discovery of the *Titanic* was no accident. The discovery fol-

Figure 1. The RMS Titanic sank in the early morning of April 15, 1912 in the North Atlantic Ocean, resulting in the deaths of more than 1,500 people. It was one of the deadliest peacetime marine disasters in history. Image courtesy of Shutterstock.

Figure 2. Photo of Dr. Robert Ballard aboard his ship E/V Nautilus in front of ROV Hercules. Ballard is the man who found Titanic, Bismarck, USS Yorktown, and John F. Kennedy's PT-109. He is passionate about the importance of exploring our oceans. Photo courtesy of Ocean Exploration Trust.

lowed careful search patterns of ocean currents and the ship's debris field. It was driven by a passion and curiosity that began in childhood to explore the ocean depths. Some discoveries do happen by accident. Discoveries made by accident are newsworthy events that often capture the headlines. The accidental discoveries of penicillin, the microwave oven, Teflon, X-rays, Velcro, rubber, and plastic are often featured by the popular press. Discoveries that are serendipitous, stumbled upon by chance, or discovered while working to discover something else, however, are rare events. Far more common are discoveries inspired by a pattern that piqued someone's curiosity.

If it happens once, it's an accident; twice, it may be a coincidence. But if it happens three times—it's a pattern.

Author unknown

The message of this book is that everyday people can make discoveries big and small by sending their brain on a mission to find patterns in the world around us. Pay attention to what your brain is trying to tell you. A popular misconception is that to discover something new, you must be really smart or really lucky, or, in view of the new era of big data, skilled at computing larger amounts of data. Neither is necessarily true. Discovery, in fact, is in reach of anyone who is curious about the world around them. Here are a few examples:

Pattern seeker and homemaker Jean Nidetch noticed that some thin people behave differently than overweight people. Some thin people, for example, often pause between bites of food and lay their fork down; some overweight people grab their fork and hang on for dear life. Nidetch not only solved her own weight problem by changing her behavior, but she also helped millions of others by founding Weight Watchers International (Nidetch, 2009).

Near to the Australian National University stands a building that causes a passerby to take a second look (see Figure 3). Its radical design consists of rows of offset zigzag walls. One day Andrey Miroshnichenko looked at the building, and a bell went off in his brain. Its radical design pattern inspired Miroshnichenko to create a new type of computer chip using light. The zigzag chip structure prevents light from travelling through its center. Instead, light is channeled to the edges of the material, permitting light to bend around corners. This discovery helps overcome a major hurdle in the goal to de-

Figure 3. Near to the Australian National University stands a building that causes a passerby to take a second look. Its radical design consists of rows of offset zigzag walls. Its radical design pattern was the inspiration to create a new type of computer chip that uses light. Image courtesy of Shutterstock.

velop optical computers capable of processing data at the speed of light.

Fish angler Laurie Rapala discovered that predatory fish repeatedly target prey that have a flaw in their swimming motion. He decided to fashion an artificial lure that mimicked the movement of an injured minnow. It worked to lure and hook record-sized fish. Word spread quickly. The Rapala lure is now the world's most popular lure, used by anglers worldwide. The discovery is also a truly great rag-to-riches story.

Guy Stewart Callendar took up climate study as an amateur enthusiast. Callendar painstakingly collected and sorted out temperature data recorded from around the world. When he evaluated the numbers, he found something startling. The pattern of numbers told him that global temperatures were rising. In 1938, Callendar was the first to discover that the planet had warmed. His work is still relatively unknown, but his contribution to climate science today was groundbreaking.

A Dutch schoolteacher discovered a new class of cosmic objects through a project that allows the public to take part in astronomy research online. Hanny van Arkel joined a team of other amateurs and professional scientists to scan the skies for black holes. One day, she put a note on the message board, and wondered if anyone knew what the bright green cluster she observed might be. The never-before-seen *voorwerp (object)* was indeed unique—it captured the moment that a once-active galaxy was dying and entering its last days (Courtland, 2008).

Just getting the right word pattern for a joke takes time. According to popular comedian Jerry Seinfeld, it can take a long time. In an interview with the *New York Times*, Seinfeld described how he created a joke featuring his first encounter with Pop-Tarts (Seinfeld, 2012). "Normally, jokes take a few days to write, but the Pop-Tart joke was rewritten several times over two years." Seinfeld claims that to achieve the right sentence structure is much like writing lyrics for a song. Instead of matching words to a melody, words need to fit the timing and rhythm pattern of the joke teller.

Homemaker Polly Murray recognized many patterns about the disease she and her family endured for many years. Interestingly, 39 adults and children living in the very same neighborhoods of Lyme, Connecticut, shared the same symptoms as she and her family: pain, redness, and rashes characteristic of rheumatoid arthritis (Murray, 1996). Her persistence to identify the cause and cure for the disease led to the discovery by Dr. Allen Steere that an insect was the vector for the disease. The disease was coined Lyme disease, a disease now present in every state.

New England Patriots Coach Bill Belichick always seems to be one step ahead of everyone else in the NFL, now one of the most winning coaches in NFL history (Halberstam, 2005). It started with his father. "Bill Belichick's father was revered for his scouting ability to recognize patterns and spot things no one else noticed, and he compiled his thoughts into lengthy reports that he delivered each week to his head coach. This also became his son's introduction to the game within the game. His father's influence remained, and Belichick has never stopped searching for winning patterns (Brown, 2015)."

Patterns waiting to be discovered can range from ones hidden in numbers like those that inspired the discovery of climate change, visual ones that revealed a common connection among people with Lyme disease, and natural patterns that underlay Laurie Rapala's motivation to invent a new kind of fish lure. The previous pattern seekers were all amateurs curious about the world around them. They were "pattern smart," but in different ways. Jean Nidetch was people smart and recognized that "we ourselves hold the instrument that makes us fat" (Nidetch, 2009). Getting the just right word pattern for a joke can take time. Comedian Jerry Seinfeld can sometimes take weeks to fit the words to the timing and rhythm pattern of the joke teller. It is Seinfeld's command of word patterns that make him one of the world's funniest comedians.

Readers of this book learn from various examples what patterns are, how diverse they are, and what we can learn from them. Because we all are gifted with the brain's innate ability to recognize patterns, the book describes how readers can recognize this skill in themselves and learn to see patterns that may have gone unnoticed with the unaided eye. The message here is to pay attention to what your brain is trying to tell you. You may discover something that is right in front of you or identify a pattern through an event that you experience every day. Sometimes, a newly developed instrument yields a new way of looking at the world. Like a new instrument, new ways to observe the world may change the way we interpret the world.

A new point of view is worth 80 IQ points.

Alan Kay, *Computer Scientist*

This quote is a reminder that looking at things in new ways can enhance one's understanding of the world around us (Stone, 2013). This book, *See the World Through Patterns*, prepares you to understand the role that pattern recognition has played in past discoveries and innovations and its significance to you, your home, and your workplace. The discoverers of the patterns provided in these examples, such as Robert Ballard, Gordon Moore, and Allen Steere, share their insights into the discovery process and the role pattern recognition played in their own breakthroughs.

   You will find clues to what distinguishes pattern discoverers from others and how pattern recognition can be nurtured. Because the brain is wired to recognize patterns, everyone has the potential to be pattern smart. A purpose of this book is to help readers become self-informed about how they are pattern smart. Are you better at recognizing people or number patterns? Do you see patterns in nature that others overlook? Do you have a gift for making people laugh? Ideally you can use the information gathered from the stories of current and past pattern seekers to apply to your own life or workplace. You may well be rewarded with an idea that could make the workplace more productive, improve the quality of human lives, or create the next new invention or product. It might even make you smarter!

CHAPTER 1

DIFFERENT WAYS TO BE PATTERN SMART

The real voyage of discovery is not seeing new landscapes, but in having new eyes.

- Marcel Proust

How is your pattern recognition skill? Do you often see how the dots are connected but overlook their importance? Have you ever tried looking at the world you know through new eyes? Do you ever see the unusual in the ordinary? Do you recognize some patterns better than others do? Do you have a passion, secret or otherwise, to discover something new? Are you looking for new ways to better your life, family, job, or the world? Are you sometimes amazed how some people who are observing the same view as you see an opportunity that you overlooked? "Patterns impose themselves on us. We must open our eyes to see them" (Horowitz, 2013).

Overview of Different Kinds of Pattern Seekers

Do you know your strengths for recognizing patterns? You will find surveys that introduce each of the five different kinds of pattern seekers in Chapter 5. The type of patterns that you will have a choice of evaluating in order are:

- People Patterns
- Number Patterns
- Nature Patterns
- Word Patterns
- Visual Patterns

Stories That Can Be Told From Looking Into Fish Ears

I have been fascinated by nature since I was a child. I am probably a good candidate to be classified as "nature smart." It likely influenced my professional career choice and shaped some of the research studies I chose to do. One study required that I look into fish ears (yes, fish have ears). When I first looked into a fish ear in 1972, I saw rings (Barkman, 1978). The concentric rings, which repeated much like a tree cross section, were etched on a structure called an otolith. Like the inner ear of humans, the otolith plays a role in hearing and balance in fish. Rings that signal the annual age of an adult had been discovered years earlier in otoliths. But *this* was new.

The rings I discovered represented the age *in days* of a week-old larvae just a few millimeters in length, about the

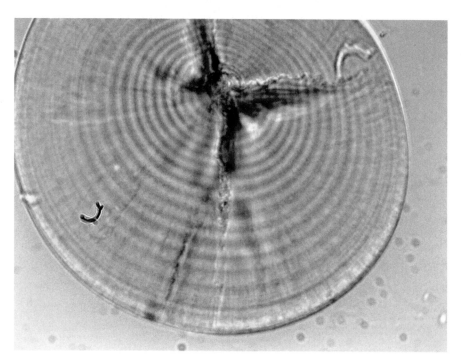

Figure 4. Fish otolith rings. I learned that rings discovered in a larval fish otolith signal the daily age of larval fish. Image courtesy of Wikimedia Commons and Edward Houde.

size of a thumbnail. The otolith itself is about the size of a pinhead. Rings continue to be laid down daily up to the age of 6 months.

A recent discovery reveals that each otolith ring even contains a fingerprint of the water chemistry of wherever the fish swam on a given day. Now, scientists hope that these rings may reveal a fish's diary of migration, from spawning through early development. The messages within a fish's ear-bone diary have implications beyond the laboratory, especially in settling sometimes hotly contested international fisheries disputes and in establishing marine protected areas (Trivedi, 2002). Since 1972, my discovery has been repeated several times and documented in many different species of fish (Fisheries and Oceans Canada, 2013).

Not by Accident

The rings were not an accident, but a pattern that inspired the discovery of new ideas about the early life history of fish. The field of otolith research has experienced phenomenal growth since the early 1970s, and it now forms the basis for hundreds of studies of early life history, age, growth, recruitment, mortality, and stock structure (Stevenson, 1992). While the field continues to grow and evolve, the otolith ring pattern will become an important tool to understanding the life habits of freshwater and marine fish life (Stevenson, 1992).

CHAPTER 2

THE POWER OF PATTERNS

Genius is seeing patterns where others see only chaos.

Erika Anderson (2013)

Patterns are powerful. They set up expectations, make connections, and inspire burning questions. They can be events that regularly repeat themselves, trends in which events rise or fall over a prolonged period, relationships that create new connections, or they can emerge from seeing the larger picture. They can be events that fall outside the norm that perhaps signal an underlying pattern. Structures that recur at different scales are called *fractals*; these are rather newly defined patterns. Seeing patterns can be life changing.

For many years, I led teachers as part of their professional development on nature field trips to introduce them to ways they could integrate outdoor study in the classroom. For some, this was a highly uncomfortable experience. The diversity of wildlife and the perceived challenge of knowing it all was overwhelming. Instead of trying to identify the wildlife, I guided them to finding patterns in nature and later challenged them to find patterns of their own. Observing lichens growing in circles, the meander of streams, and the Sun's changing position at sunrise piqued the curiosity of and sparked questions from even the most reticent participants. It created a "need to know" that favored knowing the right question to ask over knowing the right answer.

When people seek and recognize patterns in the world around them, they see order rather than chaos. This activity builds understanding of how the world works and gives people control over it. It was satisfying to observe the changes in confidence of teachers over time from reluctant participants to enthusiastic learners. More importantly, their enthusiasm rubbed off on their students, who became motivated to find patterns on their own.

Patterns Tell a Story

The great mystery of how cells replicate themselves faithfully to become genetically identical each time is a story of patterns. The mystery was solved with the discovery of DNA's structure that Watson and Crick decoded from the molecule's pattern of atomic arrangement. They mention almost casually in their 1953 article, "It has not escaped our attention that the structure postulated for DNA immediately suggests a possible copying mechanism for the genetic structure." This is one of the greatest understatements of all times. It explained the pattern of how dividing cells give rise to two genetically identical daughter cells (Nobelprize.org, n.d.).

The story of how DNA controls life's processes continues to unfold to the present.

"You deserve a break today." Patterns are coded messages like this one and once decoded, read like a story. Since 1970, CocaCola's advertising promoted a brand connecting fun, friends, and good times. Do you remember the 1971 commercial with the Hilltop Singers performing "I'd Like to Buy the World a Coke," or the 1979 commercial, "Have a Coke and a Smile" commercial featuring a young fan giving Pittsburgh Steeler "Mean Joe Greene" a refreshing bottle of Coca-Cola? Coke's central message is an invitation to pause, refresh with a Coca-Cola, and continue to enjoy one of life's simple pleasures. It's story of connecting happiness, wellness, and

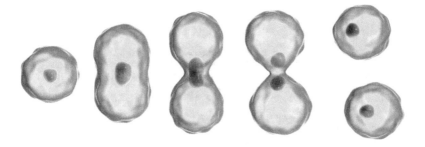

Figure 5. Dividing cell. The story of DNA explained the pattern of how dividing cells give rise to two genetically identical daughter cells. Image courtesy of Shutterstock.

Coca-Cola—has been a pattern that has repeated over many decades of successful growth (World of Coca-Cola, n.d.).

The discovery of Lyme disease reads like a detective story. A mother infected with the disease, an ecologist, and a medical detective all played a part in unraveling the cause of Lyme disease. Dr. Allen Steere had been at Yale for only four months when he saw Polly Murray of Lyme, Connecticut, infected with a disease he had never seen before. His detective work uncovered several patterns of infection that pointed to an insect cause of the disease. A few years later, an ecologist working in the field fell ill with Lyme disease after being bitten by a tick. Fortunately, he saved the tick. It was identified as a deer tick and later acknowledged to be the bacteria carrier of Lyme disease (Edlow, 2003).

Patterns Make Predictions

Patterns anticipate what's next, allowing us to make predictions of near and far future events When it was first published in 1965, Moore's Law predicted that the number of transistors that a chip could hold would double every year. Moore's Law had a far-reaching impact on technological and social development and affected many sectors of the global economy. Transistor density, even until now, has increased exponentially for the last half century. Driven by Moore's Law, new lines of laptops, smartphones, tablets, and other computing gadgets roll out constantly. Consumers continue to keep up with electronics that are faster, smaller, and more efficient than their predecessors (Li, 2013).

It is a good bet that Halley's Comet will return to Earth in 2061, 75 years after it last appeared in 1986. Edmund Halley's discovery that the comet has a periodicity of 75–76 years

dating back to 989 permits astronomers to predict the occurrence of this remarkable astronomical event far into the
future (Wikipedia, 2016, September 26).

The most powerful stroke of pattern perception is credited
to Dmitri Mendeleev when he constructed the periodic table
from data gathered from various sources. He organized the
elements known at that time into columns and rows. Because
of the design of the chart, the elements happened to fall into
families with similar properties (Thomas, 1987). When Mendeleev published the periodic chart in 1869, there were only 60
percent of the present 110 elements known at the time.

The brilliance of Mendeleev's discovered pattern is that he
predicted properties of then unknown elements that would

*Figure 6. Periodic Table. The most powerful stroke of pattern perception
is credited to Dmitri Mendeleev when he constructed the Periodic Table
from data gathered from various sources. He organized the elements
known at that time into columns and rows. Because of the design of the
chart, the elements happened to fall into families with similar properties.
Image courtesy of Shutterstock.*

later fill gaps in the original table. When the missing elements were discovered years later, his predictions were proved correct (Thomas, 1987). As Mendeleev's contribution illustrates, patterns help to create a vision of the future.

I use Google Search when I want to know, do, or buy something. More than 100 billion searches like my own are done monthly. Collectively, a record of these searches represents the pulse of our culture in real time. These searches recorded over hours, months, or years translate into trends that can become a powerful tool in the hands of marketers. Google has developed such a tool, Google Trends, to monitor searches over time (*www.google.com/trends*).

Trends in data in turn can be mined for patterns of peak buying, regional interest, and buying behavior of consumers. Together, they offer insight into what's next. Because this book fits under the genre of self-help books, I researched the past five-year trend of self-help books. The trend for the past five years has been slowly rising with peak searches recurring each year in late December to early January. The countries showing the highest interest in self-help books are the United States, India, Australia, and the United Kingdom. This kind of data, for example, offers valuable information of when and where to market the book.

Breaks in a Pattern May Reveal New Patterns (Surprises)

Malcolm Gladwell's book *Outliers: The Story of Success* featured individuals that did not fit the usual pattern. They were townspeople from Roseta, Pennsylvania; Canadian hockey players; and Korean Air pilots, who were the proverbial square pegs that tried to fit into a round hole. They stood out from the rest, and further study revealed that a sense of community plays a vital role in our health, birthdates decide the success or failure of a future hockey player, and a person's ethnic background can compromise the safety of an airplane and its passengers (Gladwell, 2008).

Outliers can also be annoying. A single data point or two that lie outside of the norm can disrupt the mean by several magnitudes and can have a deleterious effect on statistical analysis. Outliers can result from factors such as human error, sampling errors, and misreporting. Some researchers may ignore or dismiss the outlier as noise, but the wise may see a signal in the noise: An underlying new pattern.

The story of the discovery of fluoride's protection against decay began as an outlier and ended as a pattern. Dr. Frederick McKay's Colorado Springs patients were different from the rest of the population he had served. The enamel of many of his patients' teeth were mottled with a grotesque brown substance, resembling chocolate. Experts at the time blamed it on consuming too much pork, drinking bad milk, or something in the water. It was even more surprising to find that many of those patients were resistant to decay. When Colorado brown stain was found in other communities, and they too were resistant to decay, it signaled a pattern. McKay's 30-year search for the cause of Colorado brown stain revolutionized dental care, making tooth decay for the first time in history a preventable disease (McKay, 1929).

It's difficult to predict what football coach Bill Belichick will do, whether it's selecting a player in the college football draft or designing a strategy for playing defense. In his own words, "he likes to be a little bit of an outlier" (Kyed, 2016). When he joined the New England Patriots in 1970, he bucked the trend of the time of playing four players on the defensive line backed up by three linebackers. Other NFL teams soon copied

Belichick's strategy to play three defensive players on the line of scrimmage backed up by four linebackers. Belichick has a history of drafting future Patriot players overlooked by other teams in the college draft and then investing energy to develop them (Alper, 2016). The NFL's top quarterback, Tom Brady, for example, was a sixth-round pick in the college draft, and one of his favorite receivers, Julian Edelman, was drafted by the Patriots in the seventh round.

Patterns Make Connections

Patterns uncover new relationships and observing them reveals how the parts fit into the whole. People who enjoy jigsaw puzzles know that the first step in solving it is to step back and study the whole picture. This perspective offers ideas on how the pieces will fit together. Alfred Wegener illustrated this when he stepped back from a map of the world and noted how the continents fit together like pieces of a puzzle. He noticed that the coastlines of Western Europe and Africa fit together with North and South America. He also noticed that when Africa and South America were fitted together, it revealed mountain ranges, coal deposits, and fossil locations that ran uninterrupted across the continents. He wrote, "It is just as if we were to refit the torn pieces of a newspaper by matching their edges and then checking whether the lines of print ran smoothly across." This led him to conclude that the continents were once joined together in a single land mass called Pangaea. Wegener's contribution to the theory of Pangaea and continental drift is now central to the teaching of geology (Hughes, 2001).

The probability that someone will quit smoking is contingent on the type of connection a subject has with a nonsmoker (Aubrey, 2008). Among married couples, when one spouse

before   **after**

Figure 7. Pangea. The shape, size, and location of the continents before and after continental drift. Image courtesy of Shutterstock.

quits, the chances that the other spouse will quit is 68 percent. Among close friends, the probability is 43 percent. Someone's education background also made a difference. The higher the education achieved, the more likely friends are to successfully pressure each other to quit (Christakis and Fowler, 2008). The takeaway lesson from the patterns observed seems to be that it is more productive to target an individuals' connections (i.e., social network) than the individual. Alcoholics Anonymous and other institutions that practice group therapy, of course, realized this long ago.

Rachel Carson illustrated through her research that it is important to recognize that everything in nature is connected (Carson, 1962). This pattern of relationship between plants and the Earth, between plants and other plants, and between plants and animals creates a web of life that lives in a delicate balance. When even one of these connections is disturbed, life's web often ravels, leading to the death of wildlife and destruction of habitat. Unfortunately, we sometimes only understand these connections too late, when we target one member of the web to remove.

Claire D'Amour Daley would love to know what her customers want or are likely to do. Claire is vice-president of corporate communications for Big Y Supermarkets, a highly successful family-owned company based in Springfield, Massachusetts (D'Amour-Daley, 2014). She is not alone. Most organizations would love to predict their customers' actions or attitudes. Shopper scientists, such as Paco Underhill, have spent thousands of hours of field research in stores of all kinds gathering data about patterns of shopper behavior (Underhill, 1999). It's not unlike studying animal behavior. Behavior scientists study why animals do things and act in certain ways. They look at the animal's environment and try to find what caused the behavior. If the same pattern of behavior is observed time after time, it's safe to say that the behavior and the environment that triggered that behavior are connected. Research done on the human animal reveal these examples of connections.

- Research shows that there is a strong correlation between a company's growth rate and word of mouth (i.e., the percentage of their customers who are company promoters). They are the ones who say they are extremely likely to recommend the company to a friend or colleague. We already knew that, but now, research has confirmed it.
- The physical layout of a store determines how long shoppers linger and subsequently purchase products. The narrower the quarters, the less time shoppers will spend there.
- After shopping for baby diapers, new mothers often shop for camera supplies. Baskets strategically placed around

a grocery store increases sales. Social media is the number one place Millennials search to shop.

Patterns Are Never-Ending (Fractals)

*Clouds are not spheres, mountains are not cones, coastlines
are not circles, and bark is not smooth, nor does lightning
travel in a straight line.*

Benoit Mandelbrot (1983)

How would you define the shape of a cloud or a Rhode
Island coastline mathematically? Classic geometric shapes
like those used to define the world of manufactured objects—
squares, triangles, and rectangles—are not appropriate. In
fact, it is difficult to find any of these regular geometric shapes
in nature. Mathematicians have been trying to describe and
model shapes found in nature, such as clouds, fern leaves,
snowflakes, and coastlines for over 100 years. A new field of
geometry was born in the 1970s that could define fragment-
ed or irregular shapes and surfaces like clouds, coastlines,
and mountains (Fractal Foundation, n.d.). This new kind of
geometry was coined *fractal geometry* by Benoit Mandelbrot, sometimes called the "Father of Fractals" (iAwake, 2016).
Mandelbrot showed that fractals from art can be created on a
computer that mimic the creation of natural ones by repeating a simple process repeatedly.

Fractals can be thought of as never-ending patterns that
recur at progressively smaller scales; therefore, they are
self-similar whether you view them from very far away or
zoom in to view close-up.

Figure 8. Fern leaf and frond up close are an example of a fractal pattern. The smallest part of a fern leaf (frond) resembles in miniature the entire fern leaf. Image courtesy of Shutterstock.

Figure 9: Interior of La Sagrada Familia (Holy Family Church), Barcelona, Spain. Imagine the inside of the church as if it were a forest with treelike columns that divide overhead into branches to support the ceiling. Image courtesy of Wikimedia Commons.

Just as the smallest part of a fern leaf resembles in miniature the entire fern leaf, so are fractals self-similar whether you view them from close-up or very far away (Dallas, n.d.). Most coastlines, likewise, are self-similar; that is, they show the same rough, jagged irregular pattern whether viewed from an airplane or from walking along a beach's edge. Other natural objects whose smaller parts are self-like the entire object are trees, capillary beds, and lightning (Fractal Foundation, n.d.).

Why are fractals important to know about? Fractals offer innovative ways to design and engineer structures inspired by nature. There is an age-old connection, for example, between trees and architecture. Trees and their branches are common examples of natural fractals. The main branch of a tree divides into two smaller branches, and they in turn divide into two more smaller branches. This process is repeated again and again. Each new branch takes after its mother branch, creating smaller self-similar structures from the trunk to the outermost branches (Dallas, n.d.). The fractal shapes of trees offer mechanical, physical, and biological adaptations accumulated over eons of evolution. The overall shape of the tree created by fractal branching, for example, is highly resistant to wind stress. The branching pattern is the most optimal way to deliver water and other fluids to leaves, flowers, and fruit (Rian, 2014).

The relationship between tree structure and function fascinated the architect Antonio Gaudi. He often said that "there is no better structure than the trunk of a tree or a human skeleton" (Rian, 2014). He mimicked tree structure in many of his building designs coined *dendriforms*. One of his most famous dendriform designs support the ceiling of the Sagrada Familia Church located in Barcelona, Spain. Gaudi imagined the inside of the church as if it were a forest with treelike columns that divided overhead into branches to support the ceiling (Rian, 2014).

The use of branched structures enabled the use of thinner structural supports to cover large spans. Treelike structures also provide an efficient way to transfer large surface loads to a single point on the ground. It is much like waiters carrying trays above their heads. The server's fingers spread out to support the weight of the tray much like branches spread out to support the canopy. This is just one example of many of how time-tested fractals in nature inspire designs for our human-made world (Rian, 2014).

CHAPTER 3

DOING SCIENCE IS SEARCHING FOR PATTERNS

To do science is to search for repeated patterns, not simply to accumulate facts.

Robert H. MacArthur, 1984

We have learned about the world and ourselves through science. When we seek patterns in the world around us, we see order instead of chaos, which raises our confidence in our understanding of how the world works and gives us better control over it. The official Nobel Prize website, *Nobelprize. org*, is a good place to look for the role that pattern recognition plays in scientific discovery. Here you will find information for every Nobel Prize since 1901, including the Nobel

51

Figure 10: The preserved remains of ancient giant, armored, land dwelling Glyptodonts, now extinct, were similar in structure to present day armadillos. Image courtesy of Shutterstock.

Figure 11: Today's armadillo. Image courtesy of Shutterstock.

Laureates' biographies, lectures, interviews, video clips, educational games, and much more (*Nobelprize.org, n.d.*). When you plug "pattern" into the website's search engine, it returns 640 hits on the word and references to the role of patterns in research done by Nobel Prize laureates. Because pattern recognition has and continues to play such an important role in science discovery, it serves as an ideal model from which to learn and build on. Here are some stories from past giants of science that exemplify this model.

Biology—Darwin's Theory of Evolution and Natural Selection

Charles Darwin developed a scientific theory of biological evolution that explains how modern organisms evolved over long periods of time through descent from common ancestors. Darwin drew upon natural patterns of biodiversity to help synthesize his evolutionary theory from observations of life pieced together from his famous around-the-world voyage on the *HMS Beagle*. He noticed that species that are closely related but living in different habitats can vary in structure and behavior. For example, closely related tortoises living on ecologically different islands in the Galapagos varied considerably in appearance and behavior (Eldredge, 1999).

Darwin also noticed that some fossils of extinct animals were similar to living species. The preserved remains of ancient giant, armored, land-dwelling Glyptodonts, now extinct, were similar in structure to present-day armadillos.

Another pattern he observed was that species vary globally: Despite living on separate continents, distantly related species show similar adaptations to habitats that are ecologically similar. For example, the Rhea bird of South America, the Emu of Australia, and the Ostrich of Africa all are flightless ground birds adapted to the grasslands of their native countries (Eldredge, 1999).

Genetics—Mendel's Discovery of the Laws of Heredity

Our laws of heredity are a product of Gregor Mendel's careful recording of the patterns of inheritance of pea plant traits. Before Mendel's experiments, most people believed that traits in offspring resulted from a blending of the traits of each parent. However, by crossing hundreds of plants and keeping careful records, Mendel showed over and over that the offspring looked like either one of the parent plants, not a blend of the two.

This observation illustrated the dominance of one trait over another. The trait that is observed is termed *the domi-*

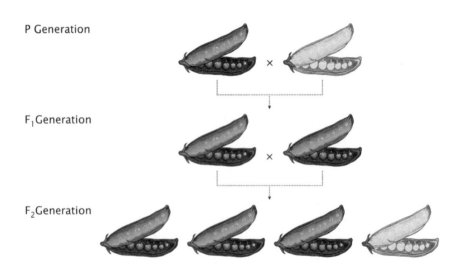

Figure 12: Mendel's Law of Heredity is a product of recording the pattern of inheritance of pea plant traits. Image courtesy of Shutterstock.

nant trait. The one that disappears is called *the recessive trait*. The Law of Dominance is just one of Mendel's three laws of heredity, which are now cornerstones of genetics (Miko, 2008).

Chemistry—Mendeleev's Creation of the Periodic Table

The periodic table is a masterpiece of organized chemical information. The evolution of chemistry's periodic table into the current form is an astonishing achievement, with major contributions from many famous chemists and other eminent scientists. A major contributor was Dmitri Mendeleev. When he arranged the elements by increasing atomic number, the chemical elements displayed a regular and repeating pattern

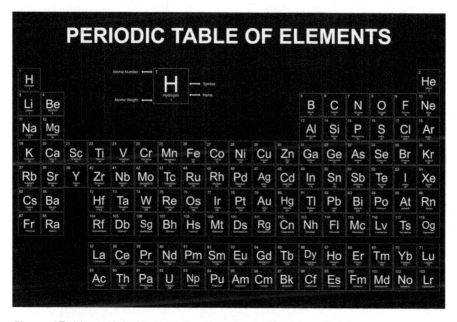

Figure 13: Mendeleev's periodic table. The table's number of new elements has grown steadily since Mendeleev's discovery of the pattern in 1869. The organization of the table has been able to explain and predict properties of elements that even Mendeleev did not anticipate. Image courtesy of Shutterstock.

of chemical and physical properties. At every eighth element, the pattern repeated much like a musical octave.

The table's number of new elements has grown steadily since Mendeleev's discovery of the pattern in 1869. The organization of the table has been able to explain and predict properties of elements that even Mendeleev did not anticipate (Judson, 1980).

Astronomy—Edmund Halley's Discovery of Halley's Comet

The expected return of Halley's Comet in 1910 created a panic when the poison cyanogen was discovered in its tail. *The New York Times* was responsible for the spread of this information after it published that the Earth was expected to pass through its tail, spending 6 hours in its vapor. The tail,

Figure 14: Halley's comet. Edmund Halley was the first to recognize the comet's periodic return to earth every 75–76 years. Image courtesy of Wikimedia Commons.

instead, missed the Earth by 400,000 kilometers, a little more than the distance between the Earth and Moon (Long, 2009). Edmund Halley was the first to recognize the comet's periodic return to Earth every 75–76 years.

During the return of the comet in 1682, Halley, by applying the laws of motion and gravitation, determined that the shape of the comet's orbit around the Sun was elliptical. When he compared its orbit to several other comets reported previously, as early as 1472, two other comets had orbital patterns identical to the 1682 one. He concluded that the comets reported in 1531, 1607, and 1682, were, in fact the same comet. Finding that its periodicity occurred every 75–76 years led him to predict that the next appearance of the comet would be in 1758. He did not live to witness it, but the comet during its on-schedule return in 1758 was named Halley's Comet in his memory (Williams, 2015).

Geology—Wegener's Discovery of Continental Drift

Until recently, geologists had thought that Earth's oceans and continents were as they seem to be now—static. They believed that little had changed since the planet formed 4.6 billion years ago. In the early 1900s, Alfred Wegener observed some interesting patterns. He noticed that the coastlines of Western Europe and Africa fit together with North and South America like a jigsaw puzzle. That similar aged plant and animal fossils were found on opposite continental shores also suggested that the continents were once joined. This visual evidence led Wegener to propose that the continents were once connected in one land mass called Pangaea. Later, Pangaea had separated over millions of years and drifted apart to become the continents we know today. Today, we explain the movements of the continents by the theory of continental drift (Hughes, 2001).

Engineering and Technology—Moore's Law

This sequence of numbers (60, 120, 240, 480, 960...) begins a well-known sequence, coined Moore's law. Named after Gordon Moore, the law predicts the number of components that can be placed on a computer chip over a certain period.

Beginning with 60 components in 1965, Moore estimated the numbers of components 10 years later in 1975 by connecting five points on a graph to form a straight line. From these data points, he projected that the number of components per

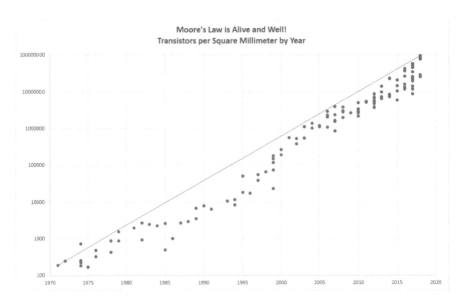

Figure 15: Moore's Law. Named after Gordon Moore, the law predicts the number of components that can be placed on a computer chip over a certain period. Image courtesy of Wikimedia.

chip would reach 65,000 in 1975; a doubling every 12 months (Fairchild Camera & Instrument Corporation, 2007). Not only was his prediction correct, computer chip production has continued to follow Moore's Law up to the present.

CHAPTER 4

SEEKING PATTERNS IS WHAT THE BRAIN DOES BEST

Over centuries of evolution, humans' pattern recognition skills determined natural selection. Hunters skilled at spotting prey and predator and telling poisonous plants from healthy ones offered them a better chance of survival than those blind to the patterns. It enabled the survivors to pass on those pattern-friendly genes to future generations.

Neil deGrasse Tyson, 2015

While the business of science is to search for patterns, it is the business of all of us to be pattern seekers. What distinguishes us from most of the rest of the animal kingdom is the desire to find structure in the information coming our way. In fact, we can't help it. Our brain craves patterns (Bor, 2012). The talent to recognize patterns is something most people don't know they need or realize that they already have. If we can turn data into a pattern or rule, then according to Daniel Bor, "near magical results ensue. We no longer need to remember a mountain of data; we need to only recall one simple law" (Bor, 2012).

A special layer of the brain found only in mammals is responsible. It is called the neocortex, the outermost layer of the brain. Because of its numerous folds, the neocortex accounts for 80 percent of the weight of the human brain. According to Ray Kurzweil, the basic structure of the neocortex is organized around groups of neurons called pattern recognizers estimated to number 300 million. Over a course of a lifetime, Kurzweil proposed that these recognizers can rewire themselves to one another to account for the new people, number, nature, word, and visual. We learn over time (Kurzweil, 2012).

We all have an innate ability to recognize patterns. But why, Brendon McConnell asks, should we use humans and not machines to find patterns (McConnell, 2013)? If we have the technology to capture data, why don't we have the technology to analyze it? Answer: A computer algorithm has not yet been developed that outperforms the human mind to analyze data and detect patterns. Becoming self-aware of our own power to detect patterns, and better skilled in how to harness it, puts us in reach of discovery and innovation. The stories of Lee Parker, Ray Kroc, and Aidan Dwyer show how everyday people can make discoveries big and small.

A new backpack sitting on top of a metal trashcan caught the attention of a homeless man, Lee Parker. Why would anyone trash a new backpack? It did not fit the usual pattern. Reminded by the remark that "if you see something, say something," he looked inside. His discovery saved many lives. Inside were five bombs planted by the alleged Chelsea bomber of New Jersey. Parker called police authorities to disarm the bomb. The world now knows Lee Parker, later cited as good Samaritan hero (Hensley, 2016).

It was common for multimixer salesman Ray Kroc to receive orders for one, maybe two, multimixers for drugstores to prepare milkshakes. But when he received an order for 10 mixers from a restaurant in California, the outlier caught his attention. Why would a restaurant need that many? His curiosity got the best of him. He traveled to San Bernardino, California, to meet the owners of the restaurant, Dick and Mac McDonald. Kroc discovered a restaurant that attracted hundreds of patrons each day, that filled orders in 15 seconds, and that made 20,000 milkshakes monthly. Kroc declared that he had to become involved in this. His discovery helped Kroc launch the McDonald's restaurants with the famous golden arches, one of the most successful fast food restaurant chains in the world (Love, 1986).

Aidan Dwyer, a seventh grader at the time, noticed something strange about the arrangements of tree branches while on a hiking trip. The leaves of the most of the tree branches he examined formed spiral patterns around the branches.

He thought that the spiral pattern was, perhaps, designed to efficiently collect sunlight. It inspired Dwyer to test his idea by building a model to mimic the structure of the tree. Instead of leaves, the natural collectors of solar energy, he fitted the model with solar panels. The tree design produced 20 percent more electricity and collected 2.5 times more hours of sunlight during the day compared to the same solar panels laid flat on a surface. Not only did his idea help him win a prestigious national science contest, Dwyer's work offers promise

Figure 16: Phyllotaxis. Phyllotaxis is the arrangement of leaves on a plant. Leaves that spiral around a stem form a distinct class of patterns in nature. Image courtesy of Wikimedia Commons.

and will inspire others to explore new ways to improve the capture efficiency of solar panels (Nguyen, 2011).

> *If I have seen further, it is by standing on the shoulders of giants.*

Isaac Newton

This familiar metaphor is attributed to Isaac Newton in a letter to his rival Robert Hooke, in 1676. Acclaimed to be one of the most influential scientists in history, Newton credited the work of Copernicus, Kepler, Galileo, and Descartes for laying the foundation of his view of the world. The metaphor nicely illustrates the progress of science. Each new generation of scientists takes our knowledge a little further by building on the results of its forebears. There is much to learn about the role of pattern recognition from standing on their shoulders. We can see further by doing it.

Fields other than science are also ripe for new discovery and innovation. Business, government, nonprofits, economics, literature, education, arts, sports, and social studies offer new fields to explore. Discoveries, both large and small, are in reach of all of us.

Intelligence and Pattern Recognition

The best thing we have going for us is our intelligence, especially pattern recognition, sharpened over eons of evolution.

Astrophysicist Neil deGrasse Tyson, 2015

Does the ability to recognize patterns define how smart you are? Often it does, but the definition of intelligence is difficult to get one's head around. Intelligence, according to the experts, is often divided into two kinds: crystalline and fluid. *Crystallized intelligence* is the kind that contestants on Jeopardy draw on to answer factual-type questions. It is intelligence that is nurtured and that grows through experience and acquiring new knowledge. As we age and accumulate new knowledge and understanding, crystallized intelligence grows. *Fluid intelligence* measures how well you can "think on your feet" by being able to respond to new situations. It is the ability to reason, learn, see connections, recognize new relationships, and find the underlying cause of things (Sternberg, 2008).

These two kinds of intelligence may be compared by considering how both are used to plan a trip. Someone may have learned through experience the skill of using Google maps to route a trip, but that same person then needs to draw on reasoning skills to sort out which of three routes suggested by Google to take. Travel time, distance, detours, traffic, and road conditions might all have to be evaluated. The navigation skill acquired to be able to use Google maps is an example of crystalline intelligence; the reasoning skills required to decide on one best route from the three that were suggested is an example of fluid intelligence.

Most intelligence tests or intelligence quotient (IQ) tests measure fluid intelligence (Hurley, 2012). According to experts, the cognitive skills that fluid intelligence mostly measure is pattern recognition. A survey of intelligence-test questions seems to support this. Anyone who has taken an intelligence test has seen matrices like those used in the Raven's Progressive Matrices: three columns vs. three rows, with three different graphic items in each row, made up of squares, circles, diamonds, or the like. One of the nine items is missing from the matrix, and the challenge is to complete the matrix by finding the underlying patterns—up, down, and across (Hurley, 2012). Do the darkened parts of the graphic shapes change simultaneously from left to right? Do the darkened parts follow the same sequence? This is the kind of reasoning that is needed to identify the missing image and correctly complete the pattern.

IQ tests are also designed to find patterns in ideas, words, symbols, numbers, and images. Pattern recognition, according to the IQ test designers, is a key determinant of a person's potential to think logically, verbally, numerically, and spatially. Compared to all mental abilities, pattern recognition is said to have the highest correlation with the so-called general intelligence factor (Kurzweil, 2012; *I Q Test Labs, 2015*). The ability to spot existing or emerging patterns is one of the most (if not the most) critical skills in intelligent decision making, though we are mostly unaware that we do it all the time (Miemis, 2010).

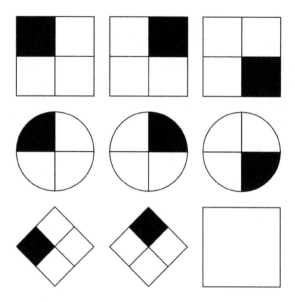

Figure 17: Raven's Progressive Matrices. It is usually a 60-item test used in measuring abstract reasoning and regarded as a nonverbal estimate of fluid intelligence. Image courtesy of Wikimedia Commons.

Measuring intelligence through matrices may seem arbitrary, but consider how central pattern recognition is to success in life. "If you're going to find buried treasure in baseball statistics to give your team an edge by signing players unappreciated by others, you'd better be good at matrices. If you want to exploit cycles in the stock market, or find a legal precedent in 10 cases, or for that matter, if you need to understand a woolly mammoth's nature to trap, kill, and eat its prey—you're essentially using the same cognitive skills tested by matrices" (Hurley, 2012).

The relationship between pattern recognition and general intelligence is reinforced by how people best learn. Expert learners recognize patterns and features that go unnoticed by novice learners. Experts, for example, "chunk" information in their domain around big ideas. History students may find it important to know the dates and places of the major wars of the 19th century, but they fail to recognize what wars have in common. Expert learners on the other hand may recognize that an underlying cause of all wars is the motivation for power and political control over others. Research on expertise suggests the importance of providing students with experiences that enhance their ability to recognize meaningful patterns of information (National Academy of Science, 1999).

More Than One Way to Be (Pattern) Smart

It is not how smart you are, but how you are smart.

Howard Gardner, 1983

Howard Gardner in *Frames of Mind* challenged the idea that there is just one kind of intelligence (Gardner, 1983). Instead of just one kind of intelligence for which one size fits

Figure 18: Multiple Intelligences. The theory of multiple intelligences differentiates human intelligence into specific "modalities," rather than seeing intelligence as dominated by a single general ability. Image courtesy of Wikimedia Commons.

all, Gardner argued that there was more than one way to be smart. There were, according to Gardner, at least eight different intelligences or IQs.

Gardener claimed that, "It is not how smart you are, but *how* you are smart." The five different intelligences that are the focus of this book are:

1. People (interpersonal) smart—the ability to recognize patterns of behavior and to understand and work with others;

2. Numbers (mathematical) smart—the ability to think through steps and sequences, and to categorize;

3. Nature (naturalist) smart—the ability to find patterns in nature;

4. Word (verbal) smart—the ability to use and understand words and how they fit together; and

5. Picture (visual/spatial) smart—the ability to draw, visualize, and orientate one's self in space.

You are happiest and most successful when you learn, develop, and work in ways that make best use of your natural intelligences. The trick to creativity, if there is a single useful thing to say about it, is to identify your own peculiar talent and then to settle down to work with it for a good long time. Everyone has an aptitude for something. The trick is to recognize it, to honor it, to work with it. This is where creativity starts.

Stephen Jay Gould

In practice, these different intelligences are not separate, but are integrated in unique combinations that shape the mind, body, and spirit of each of us. We recruit all of them daily to discover and interpret the world around us. But, because of genetics and/or environmental reasons, each of us favors or excels at one or more "smarts" over the others. Architects excel at the visual intelligence, musicians at the music intelligence, teachers at the interpersonal intelligence, writers at the verbal, and so on. Many tests exist to assess one's smarts. Follow this link to try Queendom's Multiple Intelligences & Learning Style Test: *www.queendom.com/tests/ access_page/index.htm?idRegTest=3104*

Often, however, tests are not needed. The ability to excel at one or more intelligences often emerges during the school years. Math, for example, comes easy to some students, writing to others, and leadership to others. Because Gardner was a psychologist by training, the intended audience for *Frames of Mind* were psychologists (Gardner, 1983). But educators were interested in applying the theory because they saw a fit to the classroom.

I put these different intelligences to work during an 18-day field trip with six students to study the rain forests of Costa Rica. At the end of each day, the students and I held what we coined a "sponge" session to share the observations we had individually soaked up during the day. Recognizing that the enormity of information was too much for any one person to absorb, we used the power of the group to learn from and teach each other. By pooling our different intelligences, experiences, and interests to interpret the observations, we discovered the fascinating ecology of the Costa Rican rain forests more deeply together than we could have alone.

During one sponge session, Rob was eager to share his experiences. Rob was a former gymnast who favored learning by doing. He shared that he used his kinesthetic skills to shimmy up a 100-foot herbaceous vine to reach the canopy of a *Kapok* tree. When he stuck his head through the shade of the leaf layer into the sunshine, he was amazed to find a hot dry wind blowing briskly across the canopy top, a sharp contrast to the dark, damp environment that existed just a few centimeters below the canopy. He was intrigued to find that the canopy separated two sharply contrasted worlds that support the richest diversity of tropical wildlife in the world.

Tracey was surprised to find flowers that were used as ornamentals around the home growing wild in the rain forest. Blooms that included poinsettias, hibiscus, and heleconia seemed strangely out of place in their native homes. Tracey also noticed a common feature among many of the flowers in bloom. Their petals were often colored red. It didn't take long to understand why. Because of their special ability to see red, hummingbirds, which are the jungle's chief pollinator, are wildly attracted to red.

Deb drew on her musical interest to record the noises of the rain forests. Against the backdrop screeches, howls, and bird melodies, it was impossible to identify the creatures producing the sound because of the dense foliage. It was possible to extract ways the sound patterns changed in a rhythmic way over the day and night when animals were active. We were entertained to the territorial songs of howler monkeys in the morning, the chatter of toucan birds fighting over fruit at midday, and the courtship calls of tree frogs in the evening.

Eric used his verbal skills to keep a comprehensive journal that recorded the history of each slide and transcripts of our sponge sessions. Upon our return, he told our story to a variety of enthusiastic audiences. Bob used his visual skills to document our journey on film. He took hundreds of photos to get just a few that met his high standards. He would spend up to an hour framing an image to achieve the ideal depth of focus and spatial coverage.

Brian often used the sponge session as a forum to question environmental policy and responsibility for protecting a unique environment such as the rain forest. He used his interpersonal strengths to enrich our sessions by sharing conversations he had with native Costa Ricans about their perspective on environmental issues. Despite being sympathetic to the plight of the rain forests, most native peoples whose livelihood depended on harvesting the forest for wood or clearing it for farming repeatedly resented the fact that foreigners were advising them how to manage them. It reminded us that occasionally we need to walk in the shoes of others to be aware of other sides of an issue.

These samplings from our sponge sessions illustrate that, while that all six students observed the same rain forest, each saw different patterns and interpreted them differently. All perspectives were unique, and when pieced together, created a mosaic of ecological knowledge more powerful than its parts. We all became eight times smarter.

Alexandra Horowitz decided to test the idea of how humans perceive the world around us differently in an urban environment. She invited 11 people from different walks of life such as a geologist, typographer, and naturalist, to walk with her around Manhattan and other city blocks and describe what patterns they saw (Horowitz, 2013). The geologist visualized the city as a vertical geological formation, complete with rock outcroppings, living fossils, and 60 different kinds of rocks. The typographer was attracted to the different letter fonts he saw on their walk and noted the ghost signs that revealed urban life long ago. Orb weavers, insect galls, and slug slime caught the interest of the naturalist invited to take a walk. Horowitz admits that she missed pretty much of everything that the 11 walkers observed. She blames her inattention for overlooking those things, but like all of her walkers, her search image is shaped by nature and nurture (Horowitz, 2013).

When I Received a Whack on the Side of My Head

Roger von Oech in his book, *A Whack on the Side of the Head*, advises that we all need an occasional "whack on the side of the head" to shake us out of our routine behavior and to stimulate us to ask new questions that may open up new possibilities (Oech, 2008). I received my "whack" over several years of thinking about the relationship between general intelligence (i.e., IQ) and the multiple intelligences. If we accept that much of what we call general intelligence is pattern recognition, then, does pattern recognition play an equally important role in each of Gardner's multiple intelligences? Do our perceptions of the world differ because we are pattern smart in different ways?

For example, is being numbers smart the ability to recognize number patterns that others overlook or neglect the importance of? Are people-smart individuals good at recognizing patterns in human behavior? Do ecologists perceive natural patterns that others fail to notice? Do graphic designers have the talent to recognize and create visual patterns that novices lack? Could each intelligence be enhanced through paying attention to our pattern recognition skills? I'm betting that it does.

CHAPTER 5

THE ABILITY TO DISTINGUISH PATTERNS DIFFERS FROM PERSON TO PERSON

Discovery consists of looking at the same thing as everyone else and thinking something different.

Albert Szent-Gyorgyi, 1985

I noticed early in my college-teaching career how students were smart in different ways. Some were good at numbers, others excelled at writing, and yet others were natural at leading others. I filed these observations away until I encountered the theory of multiple intelligences authored by Harvard's Howard Gardner (Gardner, 1983). Gardner suggested that the traditional notion of intelligence, based on IQ testing, was far too limited. In his theory, he proposes that there are many more ways to be smart. The theory is captured by the quote, "It's not how smart you are, but how you are smart" (Gardner, 1983). That quote redefined how I taught from the moment I learned about it. I read about it, researched it, and tested it in my classroom. I was convinced. I no longer taught as if one size fits all students but varied my teaching style to reach different kinds of learners. Thousands of teachers like me who were inspired by the same words have reinvented the way they teach.

Roses are red, violets are blue, but according to the latest understanding, these colors are really an illusion, one that you create yourself. While two people may look at the same thing, they may see things differently. You may think a rose is red, the sky is blue, and the grass is green, but it now seems that the colors you see may not always be the same as the colors that someone else sees (Robinson, 2011). It is well known that some animals see only black and white, while others can see in color. Those animals that can see in color include humans. It appears that despite the variability in the number of color receptors of the human eye, which are called cones, it is the brain that perceives colors differently.

Do You See the Patterns I See? How Our Perceptions of the World Differ

Genius is vision, often involving the gift of finding patterns where others see nothing but a chance collection of objects.

Malcolm Cowley (date unknown)

My grandkids loved to play "I Spy." It is a game I often played with them before bedtime. I Spy is a game in which the players are challenged to find different categories of objects in a picture filled with numerous kinds of objects. I am always amazed how my grandchildren are able to so quickly find the three rocking horses or the four birds, while I am still looking for the first. My grandkids see the world differently than I do. Their ability to visually recognize and group certain objects together is better than mine. Their ability to recognize certain visual patterns is well developed, even at an early age.

Not surprisingly, males and females often do not perceive the world the same way. Females, for example, are much better than males in reading someone's emotion, a key to understanding someone. When presented with faces showing happiness, fear, disgust, surprise, anger, and sadness, females heavily outscored their male counterpart in their skill to read correctly those emotional patterns (Thayer, 2000).

Our own research uncovered gender differences when 529 middle school students participated in classroom training to raise their discovery skills in science from asking the right question to identifying patterns in numbers, people, and nature. Students were tested for skill improvement after participating in a yearlong curriculum developed around activities, games, and projects and, then, retested a year later. The data showed that the students made significant gains in their confidence on these important science skills. Moreover, the results showed an unexpected difference in gender. Females were more skilled in recognizing patterns in people, while males were more confident in seeing them in natural events related to geology and weather, for example (Smist and Barkman, 1996).

Do you see a
lady's face
and the
person playing
a saxophone?

Figure 19: The well-known optical illusion of the sax player and woman challenges observers to distinguish both. Image courtesy of Shutterstock.

Why do certain people see patterns that others miss? Finding patterns, just like color, is a subjective activity. So what one person perceives as a pattern can be different from what another perceives. Our brain acts like a filter, permitting us to recognize certain patterns better than others. Similarly, the ability to differentiate patterns differs from person to person. A well-known optical illusion of a sax player and woman challenges observers to distinguish both images.

Some see only the woman, others the sax player, and yet others see both patterns. One can learn to see the other pattern, after once shown it.

Patterns are everywhere around us to discover in numbers, words, images, people, ourselves, the environment, and music. We use patterns to organize what we see and hear and to make sense of data whether we are driving in a car, listening to music, or solving mathematical problems. Different people notice different things, so what one person sees is often different from what another perceives. Unless you have trained yourself to think otherwise, it is a common mistake to implicitly assume that everyone sees and experiences the world the same way you do. They don't. There are significant implications for education, social harmony, and communication when we fail to recognize the basic truism that all external stimuli must be filtered through our individual senses, which are influenced by past events and our genetic code (Group, 2012).

Numbers Smart—Number Patterns That Inspire Discovery

A mathematician, like a painter or poet, is a maker of patterns.

G.H. Hardy, 1940

How did you learn the mathematical formula for pi? It was likely the way I learned it. I was told to commit the value pi = 3.14 to memory along with the formula to calculate the area of a circle. Savvy math teachers today encourage students to learn math by the discovery process. They coach students to learn pi the way it was discovered. Students measure the circumference of any circular object, from a tire to a Frisbee, and divide by its diameter. They discover that the number 3.14, the ratio of a circle's circumference to its diameter, recurs repeatedly. They rediscover what the Babylonians and Egyptians discovered centuries ago.

Pi was always there, a pattern just waiting to be discovered. Teaching and learning become a win-win proposition when teachers see the "aha" look of discovery on student faces. More and more of today's students learn that math is more than plugging numbers in formulas. Math is the search for patterns in numbers to reveal an underlying rule or concept (Devlin, 1998). Like many of today's students who are coached to search for patterns in numbers, numbers persons understand, practice, and even enjoy that.

Dr. Arthur Benjamin couldn't agree more. Benjamin, who is a professor at Harvey Mudd College, encourages his students to see math as fun.

He combines his passion for magic and math to show that math is more than plugging in numbers into an equation to find an answer. Away from campus, he amazes audiences around the country by performing calculations in his head faster than someone can do the calculations on a calculator (Bellano, 2019). *Reader's Digest* proclaimed him to be "America's Best Math Whiz" in their annual Best of America issue.

Figure 20: Dr. Arthur Benjamin. Called America's best math whiz, Dr. Arthur Benjamin has gained national attention for his ability to quickly perform complex computations in his head. He discovers and exploits number patterns to almost magically add, subtract, divide, and multiply numbers in his head. Image courtesy of Arthur Benjamin.

He then reveals his secrets to show how anyone can perform the same magic with numbers. His secret? He discovers and exploits number patterns to almost magically add, subtract, divide, and multiply numbers in his head (Benjamin, 2015a).

In my interview with him, Benjamin invited me to challenge him to square any 2-, 3-, or 4-digit number in his head (Benjamin, 2015b). He calculated the square of 13 faster than I could say "calculator." He then revealed how he did it. "Instead of multiplying 13 x 13 in my head, it's easier to multiply 10 x 16 (13 + 13 and 10 + 16; both add up to 26). The answer is almost 160 (10 x 16 = 160). Because 10 and 16 are both 3 away from 13, it is necessary to add 3 squared back to 160. 160 + 32 = 169." You can learn more about the secrets of mental math in his book, *The Magic of Math—Solving for x and Figuring Out Why* (Benjamin, 2015c).

Finding, describing, explaining, and using patterns to make predictions are among the most important skills in mathematics (Devlin, 1994). When perceiving a graph, for example, the first step is pattern recognition (Shah, 1997). Pattern recognition includes whether there is a straight or jagged line; if there are multiple lines; and whether lines are parallel, converging, or intersecting. In these cases, the patterns are both visually salient and conceptually meaningful. Converging lines may mean differing effects of an independent variable while parallel lines mean a matching effect. The degree of jaggedness in a line may indicate measurement uncertainty while the underlying trend is simultaneously visible (Gaukrodger, 2008).

These skills allow users of mathematics to impose order and meaning on situations that at first seem like collections of random facts. From the patterns they observe, numbers persons can predict the behavior of the stock market, land a robot on Mars, design a computer program, or craft a paper animal using the skills of origami. They enjoy and are skilled at games like Sudoku and chess. They like challenges like the following shown below.

This sequence of numbers follows a certain mathematical rule. Put your pattern thinking cap on and predict the next number after 21:

1, 1, 2, 3, 5, 8, 13, 21, ...

Did you predict "34" to continue the pattern in the sequence of numbers?

Called the Fibonacci Sequence, it begins with 1 and each number that follows is the sum of the two previous numbers (13 + 21 = 34). Leonardo of Pisa was the first to write about them over 800 years ago and named them after the pseudonym he used to author his work, Fibonacci (Garland, 1987). This sequence is notable because the ratio of any number to the next higher number or the next lower number is termed the golden ratio. Well after the first several numbers in the sequence, the ratio of any number to the next one higher is approximately .618 and to the next lower number approximately 1.1618. What is intriguing is that the golden ratios seem to appear everywhere, from defining the relationships among parts of human anatomy and architecture, to parts of plants and paintings, and even the relationship among DNA parts (Garland, 1987).

NUMBER PATTERNS ACTIVITY

Front Versus Back of the House

ENGAGE. When Danny Meyer learned for the first time in his entire career that he had more culinary school graduates working in the dining room than in the kitchen, that was the moment he said, "That practice has to stop! Our employees didn't go to cooking school to be servers" (Dubner, 2016). Culinary graduates customarily begin their careers as cooks. However, culinary grads have become outliers. They realize they can earn more money working in the front of the house as servers than cooks who work the back of the house. If you were Danny, what course of action would you recommend?

EXPLORE. "Headwaiters at top-tier restaurants can earn from $80,000 to as much as $150,000 a year including tips, according to industry executives. In comparison, a line cook might earn as little as $35,000 to $45,000 a year while working longer hours (Dizik, 2013). Some states, like New York, prevent tip sharing, which creates a pay inequity between the front and back of the house. Low wages are believed to be the biggest reason that dedicated cooks are becoming more and more difficult to find and retain (Dizik, 2013).

EXPLAIN. One solution would be increases in the minimum wage of kitchen workers paid for by raising menu prices. This would, however, create a vicious circle. The resulting uptick in menu prices, will, in turn, increase the number of tips, which will, if anything, increase the wage gap (Vettel, 2015). Moreover, it would drive away patrons.

EXTEND and EVALUATE. What solution would you suggest that would create a win-win-win solution that satisfies cooks, servers, and patrons? The answer to this could well be the answer to the same problem other establishments like Danny's have and will have in the future.

The following exercise questions whether you are numbers smart. Check off the questions for which you answer "yes." If you answered "yes" to a majority of questions, it is likely that you are especially good at recognizing number patterns.

<u>Can you/Do you</u>

- spot trends from scattered data points plotted on a graph?
- predict from a series of numbers what the next number will be?
- predict from reading an accounting of a company's or nonprofit's assets and liabilities whether it is healthy or not?
- perceive the relationship between two groups of data, like someone's height and weight?
- accurately estimate the number of open seats in a movie theater or the number of people in a crowded meeting?
- see the big picture while playing Sudoku to easily figure out missing numbers that you can plug into the right places?
- organize a random group of different numbers into groups of similar properties?

The Language of Nature Is Mathematics

I assert only that in every particular nature-study, only so much real science can be encountered as there is mathematics to be found in it.

Immanuel Kant

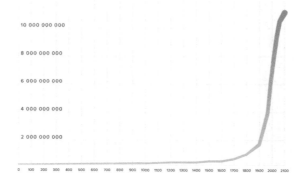

Figure 21: Trend analysis involves collecting information from multiple time periods and plotting the collected information on the horizontal line (abscissa). The objective is to find patterns from the information given. Image courtesy of Shutterstock and Wallstreet Mojo.

Figure 22: Fibonacci numbers turn up repeatedly in nature. The most common numbers of flower petals, for example, are 2 (nightshade), 3 (lily), 5 (buttercup), 8 (bloodroot), and 13 (marigold). Pine needles are found in bundles of 2, 3, or 5. The orchid with five petals is shown here. Image courtesy of Shutterstock.

These Fibonacci numbers turn up repeatedly in nature, and many can be found in your own garden (see previous sequence of numbers).

The most common numbers of flower petals, for example, are 2 (nightshade), 3 (lily), 5 (buttercup), 8 (bloodroot), and 13 (marigold). And, pine needles are found in bundles of 2, 3, or 5. The Fibonacci sequence has many applications in botany, one of which is phyllotaxis.

Phyllotaxis describes the regular repeating spiral distribution of leaves around a stem. It describes not only the leaves, but also the distribution of seeds, florets, and branches (Adam, 2009). In the 1830s, a pair of scientist brothers discovered that each new leaf on a plant stem is generated

Figure 23: Golden Angle. If you look down from above on the plant and measure the angle formed between a line drawn from the stem to the leaf and a corresponding line for the next leaf, you will find that there is generally a fixed angle between each interval of growth (Adam, 2009). This angle between successive leaves or botanical elements is close to about 137.5° (Atela, 2003). This angle is coined the golden angle. Image courtesy of Shutterstock.

at regular intervals, but not after each complete circle of a spiral. Leaves are generated every 2/5 of a circle of growth, after 3/5 of circle, or similar fractional parts of a circle. That is, if you look down from above on the plant and measure the angle formed between a line drawn from the stem to the leaf and a corresponding line for the next leaf, you will find that there is generally a fixed angle between each interval of growth (Adam, 2009). This angle between successive leaves or botanical elements is close to about 137.5° (Atela, 2003). This angle is coined the *golden angle*.

So, why has nature chosen to arrange leaves is such predictable patterns? It is not by accident. Plants have to occupy space, receive sunlight, and interact with the environment in a way that promotes their survival. Phyllotaxis is clearly a subject at the junction of botany and mathematics.

Leaves distribute themselves in mathematical ways that prevent one set of leaves from growing directly above another set leaves, which would inhibit those below from receiving sunlight and moisture. Because leaves are "socially responsible" to the plant as a whole, leaves at the bottom of a plant have an equally fair chance to survive as the uppermost ones. Can we learn from plants and adapt the mathematics of leaf display to human needs? Can we design arrangements of buildings that maximize exposure for passive solar gains? Can we create solar panels to capture maximum amount of sunshine while reducing overall area used by panels?

Seventh grader Aidan Dwyer already tried (Bells, 2011). The way leaves arrange themselves spirally on plant stems inspired Aidan to arrange solar panels in a way that could generate more energy than a flat panel. Dwyer explains in this winning essay sponsored by the American Museum of Natural History:

When I went on a winter hiking trip in the Catskill Mountains in New York, I noticed something strange about the shape of the tree branches. I thought trees were a mess of tangled branches, but I saw a pattern in the way the tree branches grew. I took photos of the branches on different types of trees, and the pattern became clearer. The branches seemed to have a spiral pattern that reached up into the sky (Dwyer, 2011).

Dwyer asked whether the purpose of the spiral patterns was to collect sunlight better. He designed experiments to test how the tree collects sunlight and built models that mimicked the pattern of tree branching

I designed and built my own test model, copying the Fibonacci pattern of an oak tree. I studied my results with the compass tool and figured out the branch angles. The pattern was about 137 degrees and the Fibonacci sequence was 2/5. Then I built a model using this pattern from PVC tubing. In place of leaves, I used PV solar panels hooked up in series that produced up to 1/2 volt, so the peak output of the model was 5 volts.

Along the way, he discovered a new way to increase the efficiency of solar panels at collecting sunlight. In his own words, Dwyer described the process:

The entire design copied the pattern of an oak tree as closely as possible. The Fibonacci tree design performed better than the flat-panel model. The tree design made 20 percent more electricity and collected 2.5 more hours of sunlight during the day. However, the most interesting results

Figure 24: The Solar-Powered Tree. In place of leaves, solar panels are placed according to the Fibonacci pattern to capture the maximum amount of sunlight. Aidan Dwyer's trees performed 50 percent better than flat panel models. Image courtesy of Shutterstock and Higyou.

were in December, when the sun was at its lowest point in the sky. The tree design made 50 percent more electricity, and the collection time of sunlight was up to 50 percent longer! (Dwyer, 2011)

A review of Dwyer's work questioned the way that he measured the electrical output of the solar panels. While the validity of the results will require repeating his experiment, Dwyer's work offers promise and will inspire others to explore ways to improve the capture efficiency of solar panels (Nguyen, 2011).

Smith College Professor James Middlebrook, who teaches architecture, thought a lot about nature and experimented with plants and animals to inspire his design compositions in graduate school. He followed the path of other designers with similar interests in recent decades. And, why not? Bio-

Figure 25: Inspired by flower designs, teachers create and construct new building structures during a professional development workshop. Image courtesy of Robert Barkman.

Figure 26: The Turning Torso residential skyscraper. The Turning Torso residential skyscraper that Santiago Calatrava designed in Malmo, Sweden, has been the tallest occupied building in Scandinavia since it opened in 2005. Like much of Calatrava's work, the aesthetic form and function of this building are closely linked. Image courtesy of Wikimedia Commons.

logical systems have been refined for specific applications for each species over millions of years through evolution. Form and function have been carefully shaped by nature's design (Middlebrook, 2015).

"The Fibonacci number sequence according to Middlebrook has long been of interest to architects, and it has informed the proportions of many buildings in previous centuries through its employment within the classical orders. There has been a long tradition within the discipline of architecture of attempting to design buildings that reflect the divine rules of nature and the universe, at least aesthetically. The refinement of mathematic laws and relationships has been paramount in this quest, beginning with the Egyptians who may have used the golden angle in the design of the Great Pyramids. The Greeks are thought by some to have based the design of the Parthenon and other structures on the golden ratio. That quest continues today" (Middlebrook, 2015).

Iranian Architect Saleh Masoumi proposes new building designs adapted to what we have discovered about leaf pattern designs to improving human living spaces. Masoumi claims that nearly all of the residential towers that have been built so far share a common weak point. Residents neither have access to the open sky nor the benefits of passive solar radiation and natural ventilation.

Masoumi envisions apartments and condominiums that cantilever in a spiral from a service core, or stem, separated from each other by a fixed angle of 137.5°, much like the phyllotaxis of leaves. Each unit is two-story, with the top level comprised of an outdoor, vegetated yard. His designs that utilize the structure like living plants provide live/work units that provide a "yard" for each individual unit (Masoumi, 2012).

It helps solve how urban living units can be both compact and people friendly. "Moreover, each unit has its own open-to-the-sky-yard, so it is possible to call each unit of these towers a house. The houses cast the possible least amount of shadow on each other due to the phyllotactic mathematical pattern, like the way that leaves in plants do" (Masoumi, 2012).

The Turning Torso residential skyscraper that Santiago Calatrava designed in Malmo, Sweden, has been the tallest occupied building in Scandinavia since it opened in 2005. Like much of Calatrava's work, the aesthetic form and function of this building are closely linked. The tower is designed with renewable energy in mind and organized around a spiral helix. As demonstrated by the number of plants and biological structures (including DNA strands) that use this geometric form, this results in a more efficient structural system.

The Eden project built in England, for example, appears much more like a plant form through its biometric appropriations of Fibonacci growth patterns. Shanghai's Natural History Museum building's shape was informed by the nautilus shell, "one of the purest geometric forms found in nature." The acclaimed architects have created 20 times the previous amount of exhibition space over six levels.

Bulls, Bears, and Numbers

If you ever followed the stock market in a bull or bear market, you will discover that neither market follows a straight curve upward or downward. Take for example the Dow Jones Industrial Average (Dow) 100-year historical chart. Each upward or downward trend is interrupted by corrections or momentary increases in the market.

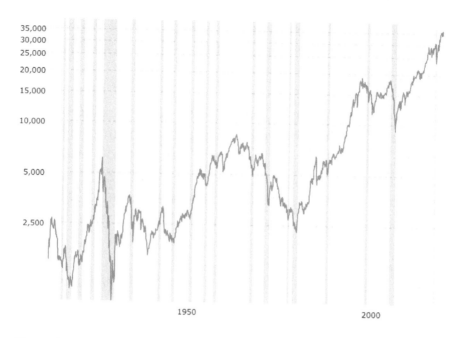

Figure 27: Dow Jones 100-year historical chart. Each upward or downward trend is interrupted by corrections or momentary increases in the market. Image courtesy of Macrotrends.net.

The science that looks for patterns in these upward and downward trends is the science of economics.

The recent bull market from 2009 to the present, for example, shows an overall upward trend, but the gains are interrupted by a series of short-lived corrections. Ralph Nelson Elliott studied these upward and downward swings in the Dow as the stock market emerged from the depression in the 1930s, looking for a pattern (Elliott, 2012). He believed that the swings back and forth reflected predictable swings in human optimism and pessimism (Garland, 1987). Elliot found a pattern or rhythm to the stock market that he coined a "wave." Moreover, Elliott discovered that investors could take advantage of the pattern to predict when to enter a market and where to get out, whether for a profit or a loss. Today, more than 80 years since his discovery, investors are still using El-liott's Wave Theory to time investments in the stock market (Deaton, 2012).

The wave principle is hierarchical. Each individual wave upon "zooming in" is in turn made up of eight smaller waves that mirror the eighth wave parent one. By "zooming out," each individual wave of the pattern is in turn comprised of eight larger waves that again mirror the parent wave. This self-similarity at different scales is the hallmark of fractal pat-terns (Cassi, 2002).

Given that the stock market unfolds in specific ways, how can investors use this information to buy or sell stock (Walker, 2010)? Once the market turns upwards (Wave 1), it is often risky and challenging to predict if the curve will unfold ac-cording to an Elliott pattern. However, when the curve reach-es Wave 4 on a correction to the market, investors would be

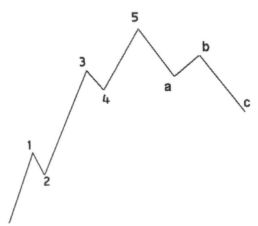

Figure 28: Elliott Wave Theory. Ralph Nelson Elliott studied the upward and downward swings in the Dow as the stock market emerged from the depression in the 1930s looking for a pattern. Elliott discovered that the market unfolded in a basic pattern of eight waves. For every five steps forward, the market takes three steps backwards. Image courtesy of Wiki-media Commons.

more confident that the curve will unfold in a way that predicts an Elliott kind of pattern. On Wave 4's pullback in the market, investors would find that it would be a good time to invest in the market. This is because Wave 5 tracks upwards and often lasts longer in duration than the other waves (Kilgore, 2015).

Using Apple as an example, here is a real-world example of the Elliott Wave pattern applied to Apple's stock movement tracked between January and April, 2015. If investors had invested in Apple stock at the bottom of correction Wave 4 ($105.00) and sold at the peak of Wave 5 ($130.00), they would have increased their investment by 20 percent (Kilgore, 2015).

Therefore, if you can identify the Elliott Wave as it unfolds and know where you are on a wave, you can predict where you are going next (Cassi, 2002). Elliott used his Wave Theory to predict the stock market rise in the middle of World War II and finance person Robert Prechter tapped it to predict the super bull market in 1982. Among market technicians, wave

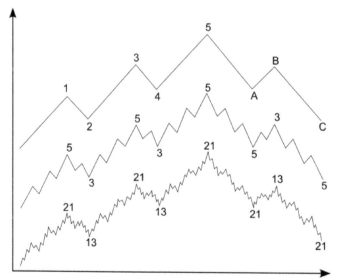

Figure 29: Elliott Wave pattern analysis of changes in Apple Stock measured between January and April, 2015. Image courtesy of Wikimedia Commons.

analysis is widely accepted as a component of their trade. El-liott's Wave Theory principle is among the methods included on the exam that analysts must pass to earn the Chartered Market Technician (CMT) designation. Robin Wilkin, Ex-Global Head of FX and Commodity Technical Strategy at JP Morgan Chase & Co., says "the Elliott Wave principle... provides a probability framework as to when to enter a particular market and where to get out, whether for a profit or a loss."

When he discovered The Wave Theory principal action of market trends, Elliott had never heard of the Fibonacci Series. However, if one examines the numbers that identify the wave parts of the cycle, one can see that there is a close connection of the cycle with the Fibonacci series of numbers: 1, 1, 2, 3, 5, 8, 13... The market unfolds in a basic pattern of eight waves. When the market is trending upwards, the market moves up and down over five waves followed by moving down, up, down in three waves. The impulse waves that lead the market trend upwards are Waves 1, 3, and 5 followed by two correc-tive waves downward. It is interesting and intriguing that the Fibonacci series of numbers that underlie the stock market are the same numbers that turn up repeatedly in nature. Are nature, bears, and bulls all predictable because of sharing a common series of numbers? It is as though we are somehow programmed by mathematics. Perhaps, we do not have as much free will as we think we have (Cassi, 2002)!

Computer on a Chip

Pattern recognition is rarely acknowledged as the thing that inspires an innovation or invention, however, new trends in technology and the explosion in digital information and tools actually have inspired recent innovations and inventions (Parmar, 2014). The ability for companies to now digitize their physical information, for example, is a new pattern. Fifteen years ago, you could have read Parmar's article only in a printed magazine; now you can read it on half a dozen different digital platforms, send it to friends, and say what you think of it via social media (Parmar, 2014).

When Gordon Moore, who co-founded Intel Corporation with Robert Noyce, described the invention of the integrated circuit (i.e., the computer chip), "there was always talk kicking around the industry for several years that someday it would be possible to create a whole computer on a chip" (Moore, 2013). This pattern of thinking gave the push to do it. "Robert Noyce, Intel's other co-founder, had the advantage of having a new semiconductor technology and cleaner technology which he combined with the need that was out there to come up with a few inventions that were necessary to go from linking individual transistors to integrating them as a circuit on a chip" (Moore, 2013). The integrated circuit can be made much smaller than a discrete circuit made from independent electronic components. The computer chip can be made very compact, having up to several billion transistors and other electronic components in an area the size of a fingernail (Wikipedia, 2014). It did not take long, according to Moore, for someone to envision the step from placing an integrated circuit on computer chip to organizing the circuits to create a computer on a chip (Moore, 2013).

Here is another number pattern to solve. What is the next number?

60, 120, 240, 480, 960...

If you answered "1920," give yourself a pat on the back. This sequence of numbers begins a well-known sequence, coined Moore's Law. Named after Gordon Moore, the law predicts the number of components that can be placed on a computer chip over a certain period.

Beginning with 60 components in 1965, Moore estimated the numbers of components 10 years later, by connecting five points on a graph to form a straight line. From these data points, he projected that the number of components per chip would reach 65,000 in 1975; a doubling every 12 months (Fairchild Camera & Instrument Corporation, 2007).

Not only was his prediction correct, but also computer chip production has continued to follow Moore's Law up to the present. Venture capitalist Steve Jurvetson described the figure illustrating Moore's Law as "the most important graph in human history" (Jurvetson, 2015). I had the opportunity to interview Dr. Gordon Moore by phone from his home in Hawaii to ask him to reflect on his and others' work that changed the future of technology. Despite being a legend in his own time, he was humble and gracious as well as quick to credit

Figure 30: Moore's Law is alive and well. Named after Gordon Moore, the law predicts the number of components that can be placed on a computer chip over a certain period. Image courtesy of Wikimedia Commons.

the work of others. In his own words, Moore describes the story behind his famous prediction.

I had the perspective that most people didn't. I was running the organization that was pursuing making increasingly complex circuits. Electronics Magazine gave me the push by asking me to contribute an article projecting the next 10 years in semiconductor components. So, I sat down with the data I had, made a wild extrapolation that turned out to be reasonably close. I had a colleague who said that should result in computer performance doubling about every 18 months. That has stuck as Moore's Law, but I plead not guilty...it was still a pretty wild extrapolation. You know the number of components that could be placed on a computer chip increased from about 60 components to more than 60,000 in 10 years. (Moore, 2013).

Moore's Law has continued unabated and celebrated its 50th anniversary in 2015 with an overall advance of a factor of roughly 2 billion. That means memory chips today store approximately 2 billion times as much data as in 1965. Alternatively, in more general terms, computer hardware today is approximately 2 billion times as powerful for the same cost (Borwein, 2015).

Can you believe billions of components per chip? Not only is Moore's Law an engine for technology growth, it also benefits the economy as a whole. As Moore first recognized, the cost of manufacturing a chip is inversely related to the increase in the number of chip components; cost decreases as the number of components per chip increases. The most noticeable effects of Moore's Law are smaller, cheaper, more energy efficient, and, of course, faster computers. In fact, each doubling of chip density results in an effective quadrupling of computational power. There is now more computing power in your personal computer than the computers that existed when NASA landed a spacecraft on the Moon.

Some say that the real importance of Moore's Law is that it serves as a business model. It gives confidence to the electronics industry to invest in the future. Because of this pattern of technological growth, industry can confidently plan and invest every year or two knowing that the electronics currently out there will become obsolete and create new demand (Hutcheson, 2005).

Global Warming

You don't need a PhD in Mathematics, however, to mine patterns in data. Up until the 1930s, most scientists thought the Earth's temperature scarcely changed. It may have been the press reports speculating that the Earth was, perhaps, warming that stimulated an Englishman, Guy Stewart Callendar, to take up climate study as an amateur enthusiast (Applegate, 2013). Callendar painstakingly collected and sorted out data recorded from around the world and found something startling. He announced that the mean global temperature had risen between 1890 and 1935, by close to half a degree Celsius (0.5°C, equal to 0.9°F). Callendar's temperature data gave him confidence to push ahead with another, bolder claim. He published a paper in 1938, claiming that carbon dioxide was responsible for the temperature rise. The data that he collected comparing carbon dioxide and temperature gave him the confidence to do so (Baff, 1999).

Date	Carbon Dioxide (ppm)
Pre-1900	290
1910	303
1922	305
1931	310
1935	320

Callendar pointed out that rising fuel combustion needed to power the industrial revolution generated 150 million tons of carbon dioxide in the first half of the 20th century and that three quarters of it still remained in the atmosphere. He connected the final dots by concluding that the agent of change responsible for the temperature–carbon dioxide connection was human. He warned that "man is now changing the composition of the atmosphere at a rate which must be very

Figure 31: The increase in carbon dioxide paralleled the rise in global temperatures measured at the same time. Image courtesy of Shutterstock.

exceptional on the geological time scale." A model that he constructed from the data he collected predicted that, if the carbon dioxide doubled, the world's temperature would increase by two degrees Celsius (Weart, 2012).

When Callendar compared the Earth's temperatures to atmospheric carbon dioxide measured at the same time, he connected the trends of Earth's rising temperature, carbon dioxide, and human fuel consumption. Coined the "Callendar Effect" at the time, concerns were expressed in both the popular and scientific press about rising sea levels, loss of habitat, and shifting agriculture zones. Today, those concerns about global warming are now real ones. The severity and frequency of storms and drought, retreat of glaciers, loss of the polar ice caps, and changes in the growing season are among the worldwide changes that we connect to global warming.

Data Mining

Now, take the recorder and fast forward from Moore's prediction in 1965 to 2016. This is the age of Big Data. We now have advanced from the five data points that Moore used to predict the future number of components that a computer chip could hold to the trillions of data points employed to discover the next drug, track down crime suspects, identify a professional baseball player, and predict the next president. People who are numbers smart and have a love of numbers have launched new types of careers called *data miners*. Instead of putting on a hard hat each day, they sit down at a computer to mine data, searching for patterns. Because of cheap storage, abundant sensors, and new software, data mining has become a multibillion-dollar industry in less than a decade.

From predicting where the next outbreak of crime might be to predicting the next flu outbreak, Big Data promises to be the next frontier in innovation, productivity, and discovery. It even will help consumers distinguish good products from bad ones. For example, one can distinguish which Honda Element cars that have fewer breakdowns by their color. When data miners crunched the numbers to compare Elements of different colors, orange had the fewest number of breakdowns (Mayer-Schonberger, 2013). Why orange? No one knows for sure, but data miners frankly don't care. Their role is to discover patterns that can be used as tools to create value in industry, medicine, law enforcement, education, science, and government. (Note: We did own a Honda Element. Unfortunately, it was green and it has had more than its share of breakdowns.)

In 2009, data miners working at Google joined with the Centers for Disease Control and Prevention (CDC) to discover new ways to prevent disease (Ginsberg, 2009). During flu season, the CDC asks physicians to report new cases of flu to help track current outbreaks and predict where future ones may occur. Because patients with flu symptoms often wait days before seeing a doctor, and it takes time for doctors to report new cases to the CDC, there is always lag time of up to two weeks between when flu patients first contract the flu and when it is reported to the CDC. Using Big Data, the engineers at Google recognized a correlation between what people searched for on the Internet and areas where there was an outbreak of flu. Google could estimate with a high degree of accuracy the current flu activity around the world based on flu patients' queries (Mayer-Schonberger, 2013). Like the CDC, they could tell where the flu had spread, but, unlike the CDC, they could tell it in real time. For epidemiologists, this was an exciting development, because early detection of a disease outbreak can predict and thus prevent its spread (Cukier, 2013).

Who wants to be a data miner? In fact, who wouldn't want to be one? If you think you have the right stuff to be a data scientist, log on to the website Kaggle (*www.kaggle.com*). Founded by Anthony Goldbloom, Kaggle is a data matchmaker, matching companies with data problems to solve to data problem solvers. He has advanced the concept of "crowdsourcing" in a big way. As of May 22, 2013, some 94,678 problem solvers work for Kaggle. They are a mix of actuaries, physicists, mathematicians, computer programmers, and engineers who hone their skills and make money competing to find the best way to mine data for patterns. The awards are significant. Prizes ranging from a few thousand dollars to a few million dollars are claimed by the data scientists who place near the top of the competition. Advanced degrees are not important, it's having the skills necessary to craft mathematical models to successfully solve data problems.

Brazilian Gilberto Titericz is one of Kaggle's top data miners and among the world's top 25 data scientists and was recently ranked number one. Titericz loves numbers and seeks to find patterns in numbers for the fun of it. If asked whether he would like to play video games, play soccer, or solve a complex math problem, Titericz would answer, "tell me about that problem" (Titericz, 2013). Very difficult data problems are extremely fun for him to solve, and he is good at them. In 2016, he was ranked second in worldwide competitions that drew 48,633 other data scientists. Participating is not easy. It requires Titericz to spend 15–20 hours a week along with balancing the demands of a full-time job and the needs of a young family. He prides himself in finishing first in a recent competition to predict the total daily incoming solar energy of 98 solar farms operated by an Oklahoma utility company.

Using data supplied by the American Meteorological Society, he created an algorithm to successfully forecast daily energy production from the Sun. Electric utility companies need accurate forecasts of solar energy production to achieve the right balance of renewable and fossil fuels available. Errors in the forecast can lead to large expenses for the utility from excess fuel consumption or emergency purchases of electricity from neighboring utilities (Titericz, 2013).

Today, amateurs of all ages, from eighth graders to adults, who have a feel for numbers and are motivated to find patterns hidden in numbers, are helping scientists overburdened with Big Data. Scientists are desperate for help from amateurs—or citizen scientists—to study, sort, or transcribe data that's already been collected and, in some cases, to even help collect data.

Unlike the challenge that Guy Stewart Callendar had of collecting data for his climate study, Vincent Granville writes that with so much data available for free everywhere, and so many open tools, he would expect to see the emergence of a new kind of analytic practitioner: the amateur data scientist. The only tool that is needed is a computer and, perhaps, a subscription to a cloud storage service. Just like the amateur astronomer, the amateur data scientist will significantly contribute to the art and science of data mining. They will eventually solve mysteries (Granville, 2013).

Could the Boston marathon bomber, Granville speculated, be found thanks to thousands of amateurs analyzing publicly available data (e.g., images, videos, and tweets) with open-source tools? After all, amateur astronomers have been able to detect exoplanets (i.e., planets outside our solar system) and much more (Granville, 2013).

You can determine if you are numbers smart by taking the test on Panamath (*http://panamath.org/index.php*). During the test, participants see a random number of circles on screen for 600 milliseconds (0.6 seconds).

Their job is to decide whether there were more yellow circles or more blue circles. Panamath measures a participant's Approximate Number System (ANS) aptitude. The simple task of deciding whether there are more blue dots or yellow dots in a brief flash says a lot about the accuracy of one's basic gut sense for numbers. Participants can view the results of their test immediately afterward and compare their performance with others in their age group.

Citizens-as-data-analysts benefit science education, and all of us because they can find answers buried deep in data that would otherwise stay buried. So put your game face on, log on to Kaggle, and compete for prizes and bragging rights. As more scientists put their data out there for citizen scientists

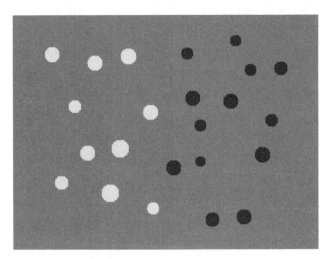

Figure 32: Panamath. Are there more yellow dots or blue dots? The simple task of deciding whether there are more blue dots or yellow dots in a brief flash says a lot about the accuracy of one's basic gut sense for numbers. Image courtesy of Panamath.

to work with, we may need to rename the "data deluge." How does "data renaissance" sound (Morris, 2010)?

People-Smart—People Patterns That Inspire Discovery

When I changed planes at an airport in Switzerland, I needed to use the bathroom. As I positioned myself over the urinal, I noticed a fly right above the drain. My male instinct was to target the fly. I am proud to say I was "right on." It turned out that the fly was not real, but was a picture adhered to the ceramic bowl. According to research studies, an object to aim for reduces spillages by 80 percent and, hence, reduces maintenance costs (Krulwich, 2009).

A Dutch maintenance worker, Jos Van Bedoff, who had served in the Dutch army in the 1960s, proposed the original fly idea years ago. As a soldier, he noticed that someone had put small, discrete red dots in the barracks urinals, which dramatically cut back on "misdirected flow." Van Bedoff discovered that the presence of a fly in a urinal literally changes the pattern of male human behavior (The Behavioural Insights Team, 2012). The use of the fly in the urinal has now spread to cities around the world and collectively has reduced spillage and maintenance costs.

Jos Van Bedoff was people-smart. He discovered a cost-saving practice by predicting patterns of male behavior. Some people, like Bedoff, are especially good at recognizing people patterns. They show that discovery is in reach of anyone who is curious, a careful observer of people patterns, and driven to understand what the pattern is telling them. The remainder of this chapter will highlight some of those people.

Figure 33: Fly in urinal. The presence of a fly in a urinal literally changes the pattern of human (male) behavior. The fly in the urinal concept has now spread to urinals in cities around the world and collectively has reduced spillage and maintenance costs. Image courtesy of Shutterstock.

Did anyone ever tell you, "I know what you are thinking?" Moreover, were you surprised to find that they were correct? Women seem to be especially good at this. My wife is one of these individuals. She is remarkably good at discovering peoples' emotions from reading patterns of facial expressions and body language that I overlook. People who excel at deciphering the meaning of human behavior patterns are people smart. People-smart individuals often share certain characteristics. They are sensitive to the feelings of others, good at understanding others, and demonstrate empathy toward others. Certain professions favor the people smart, such as the human-helping professions, which include psychology, teaching, sales, counseling, and politics. If you work in one of these people-smart professions, you might have inherited the skill to read human behavior. Even if you did not, however, it is a skill that can be learned.

PEOPLE PATTERN ACTIVITY

The Blue Marble

ENGAGE. Observe the Earth at night in the figure that follows (National Aeronautics and Space Administration, 2000). High overhead, a weather satellite circles the Earth photographing the planet every few seconds. Each satellite's small telescope scans a narrow swath of Earth's surface. It takes 312 orbits to get a clear shot of every parcel of Earth's land surface and islands. This new data is then mapped over existing Blue Marble imagery of Earth to provide a realistic view of the planet (i.e., blue marble). A composite of the Earth at night is then constructed from hundreds of the individual images. Sensitivity to light is good enough to pick up light from a 100-

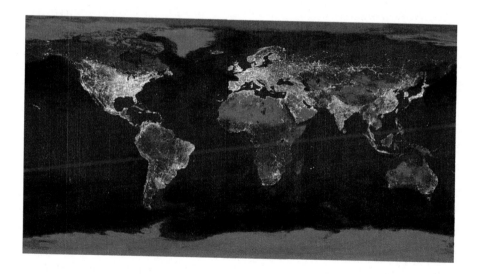

Figure 34: The Blue Marble. This image of the Earth at night is a composite assembled from data acquired by the Suomi National Polar-Orbiting Partnership (Suomi NPP) satellite over nine days in April, 2012, and thirteen days in October, 2012. It took 312 orbits and 2.5 terabytes of data to get a clear shot of every parcel of the Earth's land surface and islands. Image courtesy of National Aeronautics and Space Administration.

watt bulb! Visit this website to check out the details *https://apod.nasa.gov/apod/ap001127.html*.

EXPLORE and EXPLAIN. In daylight, our big, blue marble is all land, oceans, and clouds. The night, on the other hand, is electric. Several patterns can be recognized from the lights observed of the Earth at night. Much of the light leakage to space corresponds to street and building lights in urbanized regions (Sullivan, 1991). Find the series of lights that stretch from coast to coast along Africa's sub-Saharan savanna. These are lights from fires set to control burn tough old grasses and replace them with young shoots for cattle grazing (Croft, 1978). Notice the string of lights spanning Russia's Siberia through central China. These are the lights that string the towns and industrial centers along the Trans-Siberian and Jing-Bao/Bao-Lan (Croft, 1978).

EXTEND. The image of the Earth at night is replete with issues and potential investigations in geography, geophysics, meteorology, economics, anthropology, and environmental science (Croft, 1978). Compare the geographic distribution of populations in North America, South America, and Africa. What do the population distributions of these three continents have in common?

Contrast the amount of light produced and energy consumed by North America, Europe, and Japan to the darkness of Africa, Asia, and South America. What can you interpret about energy consumption and conservation? Why are the lighting practices of France and Germany so different from those of Belgium and England? Why is Puerto Rico so much brighter than any other Caribbean island? Other lights arise from huge burn offs of natural gas associated with oil wells. Identify these in Indonesia, Russia, Siberia, and the Middle

East. Explain the uniform distribution of tiny dots of light in the Plains states.

EVALUATE. How could understanding this map help save energy, contribute to better human health and safety, and improve our understanding of atmospheric chemistry?

The following exercise questions whether you are people smart. Check off the questions that you answer "yes." If you answered "yes" to a majority of questions, you are especially good at recognizing people patterns.

Can you/Do you

- distinguish a fake smile from a genuine smile?
- identify someone just by their gait and body posture?
- predict accurately whether someone is telling the truth?
- predict someone's job performance from their Facebook profile?
- predict from reading body language and facial expression what someone is thinking?
- project from observing a child's social interaction that a child may have special needs?
- notice that in certain professions the frequency of left-handed people is higher than right-handed people?
- recognize how and why pharmacies and supermarkets are organized the way they are?
- perceive how peer pressure influences buying habits?
- recognize that smiling for no apparent reason can make you feel happy?

I'm All Right, You're All Right

How can companies in the business of delivering packages cut costs? United Parcel Service (UPS) has an easy answer.

Listen to a UPS driver describe his route in my hometown. "We are going to make a right turn on Williams Street, then, a right turn on Laurel Street. In addition, he continues, "We are going to make another right turn on Bliss Road, then, go to end of block and make a right turn on Lynnwood Drive." Do you recognize the pattern? UPS plots its delivery routes to make as many right turns as possible. In a world where half the driving choices are left turns, they avoid turning left (Rooney, 2007).

UPS favors the right because it saves money and prevents accidents (United Parcel Service, n.d.). In an interview with Bob Stoffel, vice president of UPS, Stoffel explains that their delivery trucks rarely turn left (Shontell, 2011). The reasons are simple. When trucks turn left, they are delayed by traffic passing in the other direction. This wastes time and fuel, not to mention polluting the air. And, because it prevents trucks being sideswiped by cars traveling in the other direction, it prevents accidents. By favoring right turns, UPS delivered 350,000 additional packages in 2011 with 20.4 million fewer miles traveled. The savings each year are in the millions of dollars (Rooney, 2007; Shontell, 2011).

I'm all right, you're all right. I favor my right hand and so does most of the world's population. But asymmetry does not stop there. The data below was gathered from a high school class of 19 students (Barkman, 1991). What patterns do you see?

Body Part	Right	Left	Both
Hand	15	4	0
Eye	14	5	0
Direction	10	9	0
Leg	16	2	1
Arm	11	8	0
Ear	13	6	0
Side of Face	13	6	0
Side of Brain	2	11	6

Figure 35: Side Favored by Nineteen High School Students. Table courtesy of Robert Barkman.

Did you recognize these patterns?

- Most students are right-handed.
- If people are right-handed, they also favor their right leg and right eye.
- The side (hemisphere) of the brain favored is opposite to their favorite hand.

These patterns of laterality raise some interesting questions. Why are 90 percent of us right-handed and left-brained? What reason, if any, could account for this connection between brain and hand? Could it be an accident of nature or merely a coincidence? Or, is there something to discover from this hand-and-brain pattern? No one knows for sure, but the results of research may help to answer this question (Papadatou-Pastou, 2011).

We all communicate with our hands. To Italians, gesturing comes naturally. The people talk with their hands as much as their mouths. Isabella Poggi, a professor of psychology at Roma Tre University and an expert on gestures, has identified around 250 gestures that Italians use in everyday conversation. Some argue that hand gestures may have been the earliest form of language (Donadio, 2013).

For most of us, we are hard-wired to support this type of communication. That's because the centers for right-handedness and communication (i.e., speech and language) are near neighbors on the left side of the brain (McManus, 1999). The handedness and communication centers talk back and forth to each other to ensure that the mouth, hand, and brain are on the same page (Gentilucci, 2006). The hand gestures can help reinforce what the mouth is saying, even when two people speak different languages. If you have ever traveled in a country where English is not the native language, you likely survived the experience by relying heavily on hand gestures. Some believe that during human evolution, the skill to exercise the fine motor movements of the hands evolved before we could speak. If true, it is likely that our human ancestors talked with their hands before they uttered their first few words (Papadatou-Pastou, 2011).

In every population of the world studied so far, researchers always find a minority of left-handed people. In the United States, these outliers make up 10–13 percent of the population, but in some countries, such as China, the number is smaller because of the social pressure placed on people to be right-handed (Carter-Saltzman, 1980). Children, for example,

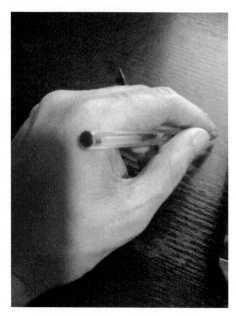

Figure 36: Left-handers make up 10–13 percent of the U.S. population, but in countries such as China, the number is smaller because of the social pressure placed on people to be right- handed. Image courtesy of Shutterstock.

are expected to hold and use chopsticks with their right hand. If, for example, a child mistakenly used his or her left hand, a father might rap the child's hand with a chopstick. Chinese schoolchildren are also threatened with slap of a ruler across the knuckles if they pick up a pen to write with their left hand (Wun, 1989).

It seems that left-handed people have been in the minority for quite a while (McManus, 1999). Studies dating back thousands of years to the time of the Neanderthals show from clues left by fossil skeletons that even a minority of our Neanderthal ancestors were left-handed. Because bones of the preferred arm are more robust, researchers can determine from a skeleton which hand is favored. University of Kansas researchers report that 89 percent of European Neanderthal fossils (16 of 18) showed clear preference for their right hands. This is very similar to the prevalence of right and left handers in modern-day human populations (University of Kansas, 2016).

Because the frequency of left-handedness has apparently remained stable over time, it suggests that there has been pressure by natural selection to preserve left-handedness in the population (Klass, 2011). But what advantage could underlie this pattern? Perhaps it is because left-handed people seem to have the advantage over right-handed people in certain areas. Left-handed people are over-represented among architects, musicians, and artists. Creativity may be a feature of left handers. Benjamin Franklin and Leonardo Da Vinci created their inventions left-handed. The works of art of Picasso and Michelangelo were created left-handed. Researchers are not sure why, but those who are left-handed seem to make up a disproportionately large part of those who are highly intelligent with IQs greater than 131. Twenty percent of all Mensa members are left-handed. When you consider that fewer than 10 percent of the total population is left-handed, this makes for many smart left-handed people (Anything Left Handed, n.d.).

A study published in the journal *Neuropsychology* suggests that left-handed people are faster at processing multiple stimuli than right-handed people (Cherbuin and Brinkman, 2006). It could mean that left-handed people have a slight advantage in sports, gaming, and other activities, such as the complex task of piloting a fighter jet in which large volumes of stimuli are thrown at someone simultaneously or in quick succession. This theory helps to explain why left-handed people excel in tennis, fencing, and boxing, as well as other sports (Panaggio, 2012; Science Daily, 2012). These and other studies suggest that by looking at someone's hands, we can learn something about the inner workings of their minds. And, it makes one wonder about all those teachers who made their left-handed students struggle to use their right hands. They may be rethinking this about now.

Brushing Away Tooth Decay

Mottled enamel is a condition in which the enamel shows grotesque brown stains as if the teeth were splotched with chocolate (McKay, 1929). At one time, mottled enamel was thought to have little relevance to the practice of dentistry. It was an outlier. Only one place in the United States was affected, and only a few practitioners would ever see this condition in a lifetime.

Frederick McKay was the first to study this anomaly after he set up his dentistry practice in Colorado Springs in 1909. There was no mention of "Colorado brown stain" in the dental literature of the day, which inspired McKay to research it himself. After trying unsuccessfully to interest other dentists in the condition, he recruited Dr. G. V. Black, a famous dental investigator to collaborate with him.

McKay and Black uncovered an interesting connection between mottled enamel and tooth decay: the teeth of children who had the condition were surprisingly resistant to decay. The mystery was solved several years later when the drinking water was tested for fluoride. In communities where patients had mottled enamel, the drinking water tested positive for high amounts of fluoride.

Grand Rapids, Michigan, became the first U.S. community to add fluoride to the drinking water in 1944. Children who were born after 1944 had a 60 percent reduction in permanent dental caries (i.e., cavities) (National Institutes of Health, 2011), proving that fluoride protects children from decay. Fluoride continues to be dental science's main weapon in

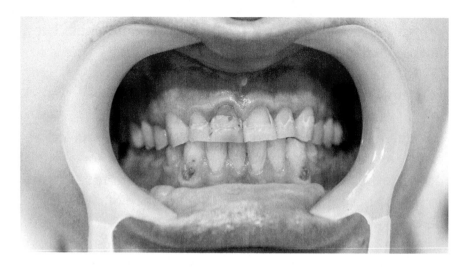

Figure 37: Children found to have Colorado brown stain (dental fluorosis) had teeth surprisingly resistant to decay. Image courtesy of Shutterstock.

Figure 38: Fish pulled through hole in the ice. The Inuit people taught Clarence Birdseye the art of ice fishing. Over a period of months observing the Inuit pull fish through holes in the ice, he noticed a pattern. The fish would thrash for a few seconds in the frozen air before instantly freezing solid. Even more striking to observe is that when the frozen fish thawed, most swam away unaffected by being frozen. Image courtesy of Shutterstock.

the battle against tooth decay. McKay's discovery that dental fluorosis protects against tooth decay ranks with other great preventive health measures of our century (National Institutes of Health, 2011).

Fresh Fish Are Always Better Than Frozen, Right?

Wrong. Modern freezing techniques make many of the fish in the freezer section superior to those in fresh seafood aisle nearby. Why? Because many fish are now frozen on the boat, just minutes after being caught, with flash-freezing units that maintain a temperature far below the typical home freezer. Many "fresh" fish are in fact previously frozen (Ducap, 2013; Szeliga, 2011; *The Any Day Gourmet*, 2013).

We can thank Clarence Birdseye's sharp eyes for patterns by inventing a way to preserve the freshness of fish. Birdseye was an eccentric inventor whose fast-freezing method revolutionized the American diet. He lived an adventurous life that included making a home in the frozen frontiers of Labrador (Kurlansky, 2012). Given that the landscape and open water was frozen for at least half the year, Labrador was a challenging place to live. Food, let alone fresh food, was a luxury. Birdseye adapted by learning how to hunt and fish (Birdseye, 1910-1916). The Inuit people taught him the art of ice fishing.

Over a period of months observing the Inuit pull fish through holes in the ice, he noticed a pattern; the fish would thrash for a few seconds in the frozen air before instantly freezing solid. He wrote in his journal that it was even more striking to observe that when the frozen fish thawed, most swam away unaffected by being frozen. To no one's surprise, the frozen fish tasted as fresh as fresh caught (Kurlansky, 2012; Birdseye, 1910–1916).

Birdseye noted that this phenomenon was limited to midwinter when the temperature was thirty degrees or more below zero. The pattern was impossible to repeat at the beginning and end of winter when winter temperatures were more moderate. Experiments that Birdseye conducted showed why. When fish were slowly frozen, ice crystals formed. The sharp points of the crystals ripped the tissues, causing tissue fluids to leak out. Birdseye later proved that "flash" freezing prevented ice crystallization, which was key to capturing the fresh taste of fish in midwinter (Fitday, n.d.; Kurlansky, 2012).

The hunger that people had for the taste of freshness had a lasting effect on Birdseye. After returning to his home in Gloucester, Massachusetts, he was hired as an assistant to the president of the U.S. Fisheries Association. His job was to find a way to deliver fish to the customer in the same condition as it landed on the docks. Until then, most fish even when refrigerated or stored on ice lost their value while being transported. Spoilage was an accepted way of life for consumers.

Birdseye was inspired by reflecting on the constant struggle for fresh food in Labrador. He realized that perishable foods could be kept perfectly preserved by the same way he had kept them in Labrador—by flash freezing. Legend has it that Birdseye spent $7.00 on salt, ice, and an electric fan, and, with these reproduced the Labrador winter. Just like in Labrador, he successfully froze fish solid in seconds. Birdseye, however, was not out to just freeze fish; he was out to create an entire frozen food industry. In September, 1922, Birds Eye flash-frozen foods was founded and continues today as an international brand of foods (Kurlansky, 2012).

On March 6, 1930, the local newspaper in Springfield, Massachusetts, ran an advertisement with the headline, "The most revolutionary idea in the history of food will be revealed today (*Springfield Union*, 1930)." Springfield was where Clarence Birdseye decided to first test market his invention of frozen foods (Kurlansky, 2012). The story claimed that frozen foods were a little short of magic. The haddock, it boasted, "was as fresh flavored as the day the fish was drawn from the cold blue waters of the North Atlantic." Consumers who had the courage to try the new kind of frozen food were pleased how much better it was than expected. Three out of four consumers of the frozen fish filets came back for more (Springfield Union, 1930).

Thanks to Birdseye's invention, a growing demand for frozen foods was predicted. Markets that could see the future and adapt to the expected popularity of frozen foods grew rapidly because of it. Founded in the 1930s on the heels of Birdseye's invention, supermarkets like Big Y of Springfield, Massachusetts, is one of those markets (D'Amour-Daley, 2014). Today, Big Y continues to study patterns of customer-buying habits much like a scientist would. Recognizing that shoppers have become more health conscience, Big Y's fish department is positioned in the flow of traffic before the meat department. A customer looking for a center-of-the-plate item will encounter the seafood department first (Bolduc, 2014).

The key to Birdseye's success was his intense curiosity. Over his lifetime, it was piqued by the people patterns he observed, from observing how the Inuit people fished and preserved their catch to recognizing the hunger that consumers had for the taste of freshness. Intrigued by finding that the freshness of fish could only be preserved in midwinter was the inspiration to discover why. Connecting these patterns over a lifetime led Clarence Birdseye to create a completely new food industry that has successfully grown into one of the world's largest.

Missing Family Found—Not by Accident

Have you ever been lost? Imagine that it is a beautiful day and you decided to take your three youngsters, ages 8, 6, and 3 years, for a short hike through a nearby national forest. You leave the trailhead about 2:00 p.m. The children wander ahead of you as you admire the wildflowers in bloom around you. While trying to keep up with your children, you lose track of which trail you took when the trail divided. Suddenly, you realize that there is not much daylight left. You round up the children and turn around to retrace your steps. But nothing looks familiar. When a hiker begins to realize that he or she has become lost, there is initially a sense of panic. As dusk approaches, your heart beats faster and you become very anxious. Many times, the first impulse is to rush back to the trail located just over the hill, around the bend, or through a stand of trees. However, this thinking seldom proves true or beneficial and typically results in the hiker moving farther from the true point of rescue. Discovering you are lost can be a frightening experience. This feeling can be compounded by fear for the children, darkness, animals, unfavorable weather conditions, and of course, death.

The above scenario is based on a true story (Rosales, 2014). Thankfully, the story had a successful conclusion. The family was discovered later that night by a search-and-rescue team. Their rescue was no accident. When setting up a search, mountain rescue teams follow certain priorities and make certain assumptions about their subjects. These assumptions are based on behavior patterns of lost subjects, in these cases, hikers (Mountain Rescue Association, 2011). An

understanding of these assumptions can help guide searchers to rescue lost hikers before their lives are threatened.

Although it is difficult to predict exactly what a lost hiker
will do when lost, searchers can infer their behavior by knowing how similar subjects have behaved in the past. Hikers take
the path of least resistance. Anything that looks like a trail is a
good idea. Hikers usually make three mistakes (Moore, 2014).
Number 1: They make a wrong turn. Number 2: Instead of
backtracking, they take another wrong turn. Number 3: Everything they do after that will get them further lost. Hunters, when compared to hikers, do not follow a trail. They are
constantly looking down at the ground searching for sign of
game and become unaware where they are. Berry pickers,
similar to hunters, focus on the ground and are often misled

*Figure 39: It is difficult to predict exactly what a lost hiker will do when lost.
Searchers can infer their behavior by knowing how similar subjects have
behaved in the past. Image courtesy of Shutterstock.*

by subtle terrain changes. Their attempt to return to familiar ground disorients them, and they become lost (Moore, 2014).

Knowing a lost hiker's age, gender, physical condition, personality traits, terrain traveled, and weather conditions, can help searchers try to walk in the hiker's shoes (Hill, 2007). The success rates are remarkably good. In a study done in Yosemite National Park from 2000 to 2010, the park responded to 2308 total search-and-rescue efforts (Doke, 2012). Of the incidents in which lost hikers were analyzed, most of the hikers were found within 24 hours. The median total time from when hikers were reported missing to when they were actually found was nine hours. The median actual search time was a remarkable two hours. Judy Moore, department chair and professor of emergency medicine at Springfield College, has had many years of experience leading search-and-rescue teams in the northeast. She reports that more than 90 percent of lost hikers are found within six hours. If the search stretches to 24 hours and beyond, the likelihood of finding lost hikers alive decreases (Moore, 2014).

The behavioral patterns constructed of rescued hikers are learned through exit interviews. The data shows that 88 percent walk downhill when lost, 73 percent find and follow a trail or path, and 82 percent are found in open areas. Based on these facts, field teams often search downhill from the last-seen point before spreading the search out in other directions (Mountain Rescue Association, 2011). If present, lost hikers tend to follow natural borders, such as streams and drainages, downhill. Watersheds, therefore, can often be used to predict the locations for lost subjects (Doke, 2012). Hikers move at an average speed of about two miles per hour, which is used to estimate how far they may have traveled since the point of last being seen. However, hikers who are in good physical shape and prepared for a variety of conditions may double this speed.

The first step in locating a lost hiker is to map the area in which the hiker was lost. Once searchers generate a map, they target where to launch the search. All of the variables that searchers must evaluate to target where the lost hiker might be found pose a challenge to even the most experienced search-and-rescue person (Moore, 2014). Researchers have recently been able to use math to help narrow down where to look (Boyle, 2010). Using a statistical model that takes many variables into account, such as terrain, behavior patterns of lost hikers, weather conditions, and time from when hiker was last seen, a map is generated to show regions where a lost hiker would most likely go. Searchers or drone-like vehicles can then cover the most probable destinations first. In a field test that simulated a search-and-rescue operation of a lost hiker (i.e., a dummy placed in the wilderness), the hiker was found in 35 minutes (Lanny Lin, 2010).

The behavior patterns of lost hikers that searchers now have in their arsenal far exceed what tools searchers had 1986. In the summer of 1986, a child, nine-year-old Andrew Warburton, became lost in the woods outside Halifax, Nova Scotia (Clugston, 2002). Within hours, a search was begun for young Andy that included more than 5,000 volunteers combing the woods, making it one of the largest ground searches in history. Despite a large-scale effort, Andy was discovered eight days too late: young Andy Warburton died in the woods. Tragedy often precipitates change. When Andy's body was finally found on the eighth day, an immediate eruption of blistering criticism hit the ground search-and-rescue team who had managed the search. No one knew how a nine-year-old would react to being lost in the woods (Clugston, 2002).

Similar mistakes were made earlier in the United States, in 1967 and again in 1968, when misfortune took the life a child who became lost while hiking in the Smoky Mountains. Search-and-rescue workers now understand some of the shortcomings of earlier search efforts. Today, using information from interviews of lost hikers who have been rescued, searchers can draw on a multitude of data from the behavior patterns of hikers lost in the woods to the type of terrain where one would expect to find them (O'Connor, 2007; Moore, 2014).

Despite the data now available to searchers, it still takes some sharp people who know how to use that data to coordinate and lead a search. Judy Moore, nationally recognized for her work, describes how leading a search is "much like detective work." She learned early about role of patterns from her dad, who often said, "'Keep looking, there is always more to see.' Finding the story behind the story was part of growing up in my home." Good advice from someone who became nationally known as an expert in search and rescue.

Patterns of Play

Why is basketball so popular? Is it because just a minimum of equipment is required to play? Is it because the rules are easy to learn? Is it because basketball can be played either inside or outside? In one of my trips to the Brazilian Amazon, I visited several small outpost villages. There were three things you could be sure to find in each rural village: a church, a soccer field, and a basketball hoop nailed to a tree. Comments from others explaining basketball's popularity were posted online. They include, "It's easy to score, that's what makes it fun." "Basketball is really good exercise and makes use of all your skills." "The lead can change rapidly and the game can be won or lost on a game-winning last shot." "Some of the greatest athletes of all times—LeBron James, Michael Jordan, Larry Bird—play or have played the sport" (The Top Tens, 2013).

The game of basketball was invented in 1891 at Springfield College (Springfield, Massachusetts), then a school that trained students for leadership positions in the U.S. and abroad in the YMCA (Bates, 1984). Basketball is now a highly popular sport (Jalan, 2014). Basketball ranks as the eighth-most-played sport in the world, and third-most-played sport nationwide (Statistic Brain, n.d.). Millions of people from youngsters to professional adults play the game year-round. It is the fifth-most-viewed sport worldwide. The NCAA Tournament ("March Madness") and the NBA playoffs attract viewers worldwide (Silverman, 2014).

How did it become so popular? Basketball draws a huge following in the United States and around the world. More than 200 basketball-playing nations compete against each other, and the team title in Olympic basketball is one of the greatest sports-related honors a country can hope to achieve (Silverman, 2014). The answer to how and why basketball became so popular is woven throughout the story of basketball's invention in 1891 by James Naismith. The story of basketball's invention is a story of patterns (Naismith, 1891).

At the time of his invention, Naismith was a professor teaching at Springfield College. In a meeting with the athletic faculty, the college's dean declared that a new competitive game was needed during the winter months to take the place of games played outdoors. Furthermore, "the game will require skill and sportsmanship, providing exercise for the whole

Figure 40: Inventor of basketball, Dr. James Naismith. The answer to how and why basketball became so popular is woven throughout the story of basketball's invention in 1891 by James Naismith. Image courtesy of Encyclopedia Brittanica.

body, yet, must be played without extreme roughness." The assignment was given to James Naismith (Bates, 1984).

The invention of basketball was not an accident. Claiming that "there is nothing new under the sun, that all games are recombinations of existing ones," Naismith studied the pattern of play of the games that existed at that time (e.g., rugby, football, soccer, and lacrosse) for their positive and negative features (Naismith, 1941). Because of possible injuries, running with the ball, which characterized outside sports, was not allowed. Instead, the rules required that players throw the ball from one player to the next. The nine player positions on the court were guided by the player positions in lacrosse (Bates, 1984). The guards occupied the defensive third of the field and were responsible for guarding the goal. The centers stood at the center third of the court and relayed the ball back and forth between the forwards and guards. The forwards were comparable to the wing players of lacrosse who were the principal goal scorers (Naismith, 1891).

A game that Naismith recalled from his childhood called "duck on a rock" managed how players would score goals. "Duck on a rock" challenged players' marksmanship by requiring players to knock off a stone (the "duck") placed on larger stone or tree stump. "A hurled stone might send the duck farther, but a stone tossed in an arc was far more accurate." In basketball, the ball, like the duck, is tossed in an arc toward the basket to score. The ball, Naismith decided, had to be soft enough not to harm the players and large enough to be seen. Soon after Naismith posted the 13 rules on the gymnasium door, and the first ball was thrown into the air, the game took on a life of its own (Naismith, 1941; Bates, 1984).

The game became instantly popular beyond Springfield. Naismith's students went on to teach at YMCAs around the country and the world, bringing with them the rules for this new game packed in their Bibles (Wolf, 2002). YMCAs around the world now had a new indoor game that could be quickly learned and implemented (Bates, 1984). Because other sports at the time were played outdoors, basketball was an outlier. Its break from the existing pattern, however, inspired a new pattern; football, soccer, and other sports soon followed basketball indoors. Much like when Alfred Wegener stepped back from a map of the world and noted how the continents fit together like pieces of a puzzle, Naismith stepped back from other sports played at the time to study what parts of each would be valuable to his new game.

Reading people patterns play a role in every sport. Malcom Gladwell describes how top tennis coach, Vic Braden, could predict when a server would default just before contacting the ball. There was something in the server's body language

Figure 41: The nine player positions on the basketball court were guided by the player positions in lacrosse. The guards occupied the defensive third of the field and were responsible for guarding the goal. The centers stood at the center third of the court and relayed the ball back and forth between the forwards and guards. Image courtesy of Springfield College's Special Collections and Library Archives.

Figure 42: The expanded original 13 rules of basketball. Soon after Naismith posted the 13 rules on the gymnasium door and the first ball was thrown into the air, the game took on a life of its own. Image courtesy of Springfield College's Special Collections and Library Archives.

that tipped Braden off. This is remarkable, given that professional tennis players only default three or four times out of 100 serves (Gladwell, 2005).

Tennis lovers always look forward to the Wimbledon tennis tournament each year. Recognized as the oldest and most prestigious tennis tournament in the world, it is the only major tournament played on grass. Unlike other surfaces, which are either hard or clay, grass shows a wear pattern from players running and sliding to reach the ball. It is curious that when the wear pattern is compared between the earlier years of tennis played on center court to recent play, there is a marked difference.

Today, the wear pattern is visible largely around the baseline. In the 1980s, the wear was visible both at the baseline and around the net. We can conclude that while yesterday's players alternated between coming to the net to volley and playing from the baseline, today's players spend most of their efforts returning ground strokes from the baseline. What caused this shift in play strategy? Is it due to changes in equipment or changes in players' skill or something else?

What Is Your Face IQ?

Is it just coincidence that Frank Sinatra and daughter, Nancy, are world-class singers? Or that 206 football players today have fathers who played in the NFL (Pro Football Hall of Fame, 2013)? Is it just coincidence that children of alcoholics are about four times more likely than the general population to develop alcohol problems (National Institute of Alcohol Abuse and Alcoholism, 2012)? Questions like these have long challenged psychologists, biologists, and philosophers. These yet-to-be-answered questions pertain to one of the world's greatest mysteries. Coined the nature-versus-nur-

Figure 43: In the 1980s, the wear pattern of the Wimbledon tennis court was visible around both the baseline and net. Image courtesy of Tennis Hall of Fame.

Figure 44: Illustration of an authentic smile (image on left) compared to a false one (image on right). Image courtesy of Dawn Barkman.

ture mystery, it questions the relative role that genes and the environment play in shaping human behavior. The interplay between genes and the environment determines everything from athleticism to a person's predisposition to alcoholism to someone's personality (Miller, 2012).

One way to appreciate the nature-versus-nurture question is to study ourselves... smiling. Find a photo or two that illustrates you smiling. Is the "say cheese" smile genuine? That is, were you smiling because you were genuinely happy? Or was the smile done for the sake of the photo? To understand the difference, compare the photos below. Which smile is authentic? (Note: You can also take an online quiz at the BBC: *www.bbc.co.uk/science/humanbody/mind/surveys/smiles.*)

Although fake smiles often look very similar to genuine smiles, they are slightly different, because they are brought about by different muscles that are controlled by different parts of the brain (Durayappah, 2010). During an authentic smile, both the lips and eyes are involved. One muscle tugs the lips upward and another squeezes the tissue around the eye into the shape of a crow's foot (Jaffe, 2010). In the "say cheese" smile, only the mouth is involved. The zygomatic major muscle, which resides in the cheek, is the muscle that is responsible for working the lips. When working alone, it produces a false smile. During an authentic smile, the orbicularis oculi muscle that surrounds the eye joins the action of the zygomatic muscle to construct a true smile. A true, spontaneous smile is sometimes called the Duchenne smile coined by Paul Ekman to honor the anatomist Duchenne de Boulogne, who first dissected the facial muscles responsible (Ekman, 1990).

The differences between smiles that express either true or false joy point to how nature and nurture interact. Fake smiles, depending on the situation or environment, can be performed at will, concealing one's true emotions. The brain signals that create them come from the conscious part of the brain and prompt the zygomatic major muscles in the cheeks to contract. This is nurture at work. Genuine smiles, on the other hand, are generated by the unconscious brain, so they are automatic. This is the reason that an authentic smile can "only be put in play by the sweet emotions of the soul" (Ekman, 1990). When people feel real pleasure, signals pass through the part of the brain that processes emotion, prompting muscles like the orbicularis to contract. Brain control of the orbicularis is shaped by genes, illustrating nature at work. The interplay between voluntary and involuntary control of the facial muscles nicely illustrates the roles that nature and nurture play to shape our facial patterns and behavior.

True joy is one of the six or seven emotions recognized by Paul Ekman. They define our lives: joy, anger, disgust, surprise, fear, and sadness (Ekman, 2003). Contempt is sometimes included to raise the total to seven emotions (Cherry, 2014). Each emotion has a unique pattern, the most identifiable patterns being in the face and voice. Paul Ekman set out to research whether the emotions were universal (i.e., shaped by nature) or culturally variable (i.e., shaped by nurture). He shared photos of each facial expression to people living in five different cultures—Chile, Argentina, Brazil, Japan, and the United States—and asked them to judge what emotion was expressed by each facial expression. The majority of each culture agreed on each emotion expressed by the photo (Ekman, 2003).

His cross-cultural study suggested that the key facial expressions were innate, shaped by nature, and universal to our species. However, his conclusion left some questions unanswered. For example, could all the people he studied have learned the meaning of facial expressions through the media or having contact with people with other cultures? Recognizing that he needed a culture to test that controlled contact with other cultures, Ekman chose to study a native tribe of Papua New Guinea who up until his visit had little contact with the outside world. When showed photos of the seven different emotions, the New Guinea natives had no trouble identifying each emotion correctly. The cross-cultural studies proved with little doubt that that it doesn't matter what culture you are from; joy, anger, disgust, surprise, fear, and sadness are shown by all people in every culture on the planet, and they show them the same way (Ekman, 2003).

The study of facial patterns led some to wonder whether reading faces offered any practical value. Could this skill, for example, be used by the FBI, TSA agents, and police groups to spot terrorists or determine if a suspect is lying? The answer is yes. Paul Ekman writes, "The face is a dual system, including expressions that are deliberately chosen and those that occur spontaneously, sometimes without the person even aware of what emerges on his own face" (Ekman, 2009). The spontaneous ones can leak a person's real feelings, exposing someone who is hiding the truth. If, for instance, you ask someone "How's it going?" and they respond "Great" when things are not, the face will exhibit a momentary expression of sadness. Unless you are a truth wizard, these micro-expressions are invisible to most people. This is because these micro-expressions happen extremely fast, lasting only 1/25 to 1/50 of a second (Ekman, 2009; *Eyes for Lies*, 2005).

Most anyone can be trained over time to detect these micro-expressions. There are, however, a few people who seem to be naturals at detecting someone lying. Renee is one of them. (Note: For her personal reasons and to preserve her anonymity, I have referred to her using her alias. Renee authors the website, Eyes for Lies at *www.eyesforlies.com*). She is one of 20,000 persons who took part in a study conducted by Paul Ekman and Maureen O'Sullivan to test a person's ability to detect deception (Edward, 2009). Out of 20,000 people studied, only 50, including Renee, were able to do so. They were called "truth wizards." O'Sullivan describes the wizards as being extraordinarily attuned to detecting the nuances of facial expressions, body language, and ways of talking and thinking. They had in common the ability to observe a video-tape and detect within seconds micro-expressions and other clues to deception of subjects who were either telling the truth or lying (O'Sullivan, 2009).

I asked Renee to describe an example of how she might put her talent to work (Renee, 2014). "If I were to meet a man in the airport, I would look at every element of him (naturally, out of curiosity to understand him). If he stands next to me and is immaculately dressed, doesn't have a whisker or hair out of place, and I can see his zippers on his carry-on are all matched and centered, it tells me I am likely dealing with a person who is a neat-freak, or, potentially an 'A' type person-ality. So, if I hear him talk to co-workers and say in the middle of his conversation, 'Oh, I lost that at home. I am so sloppy and disorganized. I'm sorry...' I am going to raise my eyebrows in disbelief. Why? Because he gives me behavioral evidence in his attire and grooming to suggest otherwise" (Renee, 2014).

Renee now consults with corporations, professionals, and law enforcement to train others to detect deception. She al-ways advises clients to judge someone who may be lying not by one red flag, but three. "Observe not only what facial ex-pressions tell you, but also what their body language and what their voice tells you." She thinks that to be able to recognize the red flags of deception, is to always seek to understand. "Ask questions. The more curious you are, the more you will learn and see and the more that you will be able to connect the dots" (Renee, 2014).

Her discovery that she had a talent to judge people patterns better than most was not a "eureka" moment. She always suspected that she had this talent, but, she did not reach out until her thirties to confirm it. That was when she was invited to join the wizard project. According to O'Sullivan who conducted the study and interviewed her, Renee, fits the model of a truth wizard perfectly. Wizards are curious, always paying attention, and highly motivated to understand others and discover what makes others tick. Heredity, perhaps, plays a role as well. Her mother is also a truth wizard.

Nature-Smart—Natural Patterns That Inspire Discovery

I'm always trying to see a pattern in a forest and I'm tickled that I can do that.

Stephen Jay Gould

New products and innovations created by mimicking patterns in nature are well known to those who have been inspired by its design (Benyus, 1997). "This is the idea behind the increasingly influential discipline of biomimicry: that we human beings, who have been trying to make things for only the blink of an evolutionary eye, have a lot to learn from the long processes of natural selection, whether it's how to make a wing more aerodynamic or a city more resilient or an electronic display more vibrant" (Vanderbilt, 2012).

In the future, you won't need to travel farther than your house to experience how patterns in nature inspired its design. The house that your children or grandchildren might live in will have likely been modeled after a termite mound, a basking snake, and a green leaf. The house of the future exists now. There are over 100 of them in the United States, competing to be certified as a "living building" (Wikipedia, 2016, October 21). The Hitchcock Center, located in Amherst, Massachusetts, is one of them. The building's architect says the requirements to become certified are rigorous (Batchelor, 2016). The building must meet seven different performance areas: site, water, energy, health, materials, equity, and beauty.

Most, if not all, of these requirements to become a living building are being met by mimicking how nature solves its energy, water, and other problems. By copying how leaves convert sunlight into electrical energy, the Hitchcock Center has totally eliminated its dependence on fossil fuels. One hundred percent of its annual electricity needs are being met through rooftop photovoltaic panels. Its living building models how nature conserves water by recycling it. The building is a watershed, capturing rainwater on sloping roofs that send water to an underground reservoir. After being filtered and subjected to ultraviolet treatment, the water is pure enough to drink. It is remarkable that the center is not tied to a municipal water supply (Batchelor, 2016).

Just like a snake that maintains a constant body temperature by moving between sun and shade, the building maximizes passive heating and cooling. The building's largest surface area faces south to absorb the Sun's rays in the winter. Shades are designed to help cool the center in the summer. Fans and windows are operated to either suck the heat out in the summer or recycle heat in the winter using design methods inspired by the self-cooling mounds of African termites (Doan, 2012).

The invention of the zipperless zipper was inspired by a troublesome weed called burdock (Mother Nature Network, 2010). Hikers often encounter the seed of burdock sticking to their clothing. George Mestral was curious about how the seed could stick and unstick to clothing when gently pulled. Mestral successfully reproduced the design of the burr seed with two pieces of fabric, one surface with tiny hooks and another surface with tiny loops. When pressed together, the two fabrics "hooked" up and unhooked when given a push and pull. We know it today as Velcro (Suddath, 2010). Hearing aids modeled after the eardrums of parasitic flies, wind turbines modeled after flippers that propel Humpback whales through the water, and concept cars modeled after fish are some of the many nature-inspired products whose ecological roots have changed our lives (Websdale, 2011).

New innovations from low-tech to high-tech ones have been inspired by nature. A revolutionary new method to encrypt confidential information was inspired by human biology (Stankovski, 2014). When the body exercises, the heart and ventilation speed up to meet the oxygen needs of the muscle. Coordination between the two systems is achieved by the complex nerve information that passes back and forth between the heart and lungs. Researchers have captured the nature of this information exchange and used it to convert text into an endless number of uncrackable codes. This discovery promises to make life tough for cyber criminals and government spy agencies (*ScienceDaily*, 2014).

Paul Sperry discovered a solution to prevent slipping on wet surfaces such as the surfaces found on his sailing vessel. It was just below his feet, or more precisely, the paws of his sure-footed cocker spaniel dog. He noticed that the pads of the paws of his cocker spaniel were imprinted with wavy-like

Figure 45: How do you cool a building without air conditioning? In nature, termites build skyscraper-like mounds that are ventilated by a complex system of tunnels. The result is an architectural marvel that achieves 90 percent passive climate control by taking cool air into the building at night and expelling heat throughout the day. Image courtesy of Wikimedia Commons.

Figure 46: Hikers often encounter a plant called Burdock that sticks and unsticks to clothing. It is the plant that inspired the fabric now known as Velcro. Image courtesy of Wikimedia Commons.

grooves, much like those found on some road surfaces to prevent cars from sliding when roadways were wet.

A light bulb went on. Sperry created a shoe with a bottom modeled after his dog's paws with the same herringbone pattern. The traction was obvious when Sperry walked on ice or any slippery surface. Today, we know the shoe as the Sperry Top-Sider (DeYoung, 2015).

People who are nature smart recognize small- and large-scale patterns. Large-scale patterns are the patterns that can be seen from a satellite or airplane. My favorite are the patterns which can be seen on the Earth at night. Astronauts circling the Earth have the wonderful vantage point of observing the nighttime Earth from 350–400 kilometers above the surface, taking in whole regions at once. Onboard cameras and a bit of experimentation allow astronauts to take highly detailed images of our cities at night and share them with the rest of us. It has helped many people visualize the world's distribution of people and cities (Simmon, 2008).

The first pattern that jumps out at you from viewing the Earth at night is the difference in light observed between the north and south hemispheres. The much brighter northern hemisphere reflects the greater distribution of people north of the equator and their higher energy usage. From a geographic perspective, cities at night tell different stories about a region. City lights provide sharp boundaries that delineate the densest concentrations of people, a characteristic that has been used to assess the effect of urbanization on Earth's ecosystems (Simmon, 2008).

In many cities, neighborhoods of different generations can be distinguished by the lighting color and patterns along their streets. In many North American cities, older neighborhoods

Figure 47: Pads of dog paws with wavy-like grooves inspired the invention of shoes with the same pattern on the bottom. It provides traction when surfaces were wet. Image courtesy of Wikimedia Commons.

have less regular street patterns and light green mercury vapor lighting, while newer cities, especially in the western United States, have street patterns aligned to the compass directions and use orange sodium vapor lighting (Simmon, 2008). The major Denver street patterns are rectilinear, aligned north-south and east-west. Cities from different regions of the Earth are also identified by differences in their nighttime lights. Japanese cities glow a cooler blue-green than other regions of the world. Newer developments along the shore of Tokyo Bay are characterized by orange sodium vapor lamps, while the majority of the urban area has light green mercury vapor lamps (Simmon, 2008).

Nature makes nothing useless.

Aristotle

The following exercise explores whether you are nature smart. Check off the questions that you answer "yes." If you answered "yes" to a majority of questions, you are especially good at recognizing patterns in nature.

Can you/Do you

- identify products designed for human use that were inspired by nature?
- notice how flowers follow the sun during the day?
- recognize different species of birds from their flight patterns?
- reconstruct the history of a woods or field from clues left from the past?
- look out the window of an airplane and wonder about the repeating circles frequenting the landscape or the serpentine nature of rivers and streams?
- notice that some of the first plants emerging in the spring

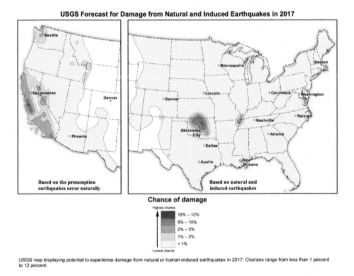

Figure 48: United States Geological Services forecast for damage from earthquakes in 2016. Oklahoma now leads California in yearly quakes. Courtesy: Mark Peterson et al, United States Geological Survey, Department of Interior.

now appear earlier than in the past?
- recognize how earthquakes now occur in places they were once rare?
- notice how dandelions show a remarkable growth spurt just before the flower turns to seed?

The Wiggle That Fish Cannot Resist

Who is the perfect predator? The perfect predator is one that removes prey from a population that is about to die anyway (Sinclair and Arcese, 1995). They are usually the old or the sick and injured. It is a win−win situation for both the predator and prey population as it keeps the prey population numbers in check and sustains the health of the predator population. Predatory fish like bass, trout, and salmon and their smaller minnow prey follow this rule. The first to recognize and understand the significance of this pattern was Finnish angler Laurie Rapala.

During the hard economic times in Finland in the 1930s, Laurie Rapala fed his family by fishing the lakes around his home. He often set up to 300 hooks at a time and waited, sometimes days, to catch his quota of fish. To occupy his time, he would observe the bass feeding on the nearby schools of minnows (i.e., small baitfish). He repeatedly observed that the predatory bass would always target a particular bait minnow. The doomed minnow invariably had a flaw in its swimming rhythm characterized by a certain wiggle and twisting motion, suggesting that the minnow was either sick or injured (Schara, n.d.).

A light bulb went on! Rapala decided to fashion a lure that mimicked the movement of the injured minnow. He carved the body first out of bark and later out of balsa wood. To imitate the silvery body scales, he wrapped the body in tinfoil scraped from the inside of candy bar wrappers and drew scales on it. He repeatedly tested it to get just the right motion by dragging it through the water as he rowed a boat. When the time came to try the new lure, the game fish wasted no time attacking it. The story has it that Rapala once landed 600 pounds of fish in one day. This was a far better alternative than baiting 300 hooks (Farmer, 1979).

Word of mouth of the lure's success traveled quickly, and people soon vied to represent the Rapala lure in North America and elsewhere. Worldwide sales rose quickly. Today, Rapala lures have been anglers' favorites since Laurie launched the business of manufacturing and selling them in the 1930s.

Figure 49: To occupy his time, Laurie Rapala would observe the bass feeding on the nearby schools of minnows (small baitfish). He repeatedly observed that the predatory bass would always target a particular bait minnow. The doomed minnow invariably had a flaw in its swimming rhythm characterized by a certain wiggle and twisting motion, suggesting that the minnow was either sick or injured. Image courtesy of Shutterstock.

Sales now reach 20 million per year (Mitchell, 2005; Schara, n.d.).

Rapala's invention spawned an entire industry to exploit the feeding behavior of sports fish prey (Cermele, 2013). Walk into any box store and browse the fishing department and see the display of tens, if not hundreds, of different kinds of fish lures. They are all designed to mimic a fish predator's natural food in the way that it swims, smells, and sounds (Jones, 1992; Kageyama, 1999; Schultz, 2012). Nevertheless, it is still Laurie Rapala's lure that anglers choose to add to their arsenal of lures.

"That's Funny"

"That's funny," remarked Alexander Fleming after noting the unusual appearance of a bacterial colony he was culturing on a petri dish. The bacteria that Fleming was culturing were a

Figure 50: Walk into any box store and browse the fishing department and see the display of tens, if not hundreds, of different kinds of fish lures inspired by a pattern that Rapala recognized. Image courtesy of Wikimedia Commons.

human pathogen of the family Staphylococcus. The petri dish that caught his attention was different from the other petri dishes he was preparing to clean (PBS, 1998). The yellow colonies of bacteria seemed to disappear around a white fluffy blob of mold (Fleming, 1929).

The clear transparent area where the bacteria had once grown reminded him of an observation made years earlier. During research of antiseptics to combat wound infections, he discovered a powerful antiseptic when his nose chanced to drip on a colony of bacteria. When fluid from his nasal secretions was cultured with bacteria, its colonies were destroyed (i.e., lysed), leaving a transparent area where the bacteria once grew (Fleming, Nobel Lecture, 1945). The agent responsible was coined lysozyme, one of our own natural defenses against infection (Goodsell, 2000). Were Fleming's discoveries really accidental as the press has described them?

The mold-destroying bacteria was also a contaminant, likely created when a microscopic spore from the mold alighted on the petri plate. Fleming identified the mold as *Penicillium chrysogenum*, a mold that thrives in moist, cool places and the source of the antibiotic penicillin. Fleming's discovery of penicillin was the world's first antibiotic and the first to be isolated from a mold (McFarlane, 1984).

Fleming had neither the laboratory resources nor the chemistry background to take the next giant steps of isolating the active ingredient of the penicillium mold juice, purifying it, figuring out which germs it was effective against, and how to use it. That task fell to Dr. Howard Florey, a professor of pathology at Oxford University (Markel, 2013). Florey and his team successfully manufactured enough penicillin from the

Figure 51: "That's funny," remarked Alexander Fleming after noting the unusual appearance of a bacterial colony he was culturing on a petri dish. The colonies of bacteria seemed to disappear around a white fluffy blob of mold. Image courtesy of Wikimedia Commons.

liquid broth in which it grew to test it on humans (Nobelprize. org, 1945).

One of the first patients to be tested was a friend of Fleming who developed meningitis. When Fleming's friend appeared to be dying, Fleming appealed to Howard Florey to bring him enough penicillin to treat his friend. The patient remarkably recovered, which became a turning point in the history of medicine (McFarlane, 1984). Since then, the discovery of penicillin changed the course of medicine. It has enabled physicians to treat formerly severe and life-threatening illnesses such as bacterial endocarditis, meningitis, pneumococcal pneumonia, gonorrhea, and syphilis. Fleming's discovery that molds such as penicillin could be mined for antibiotics spawned an entire industry. Penicillin, today, is the most widely used antibiotic in the world. Several pharmaceutical companies began screening a variety of other natural products for antibacterial activity, which led to a whole host of new antibiotics, such as streptomycin and tetracycline (Eickhoff, 2008).

The media often paints the discovery of penicillin as an accident. Even Fleming thought so. In his Nobel lecture (*Nobelprize.org*, 1945), he remarked that his discovery was inspired by a chance observation. However, pattern recognition, not chance, played a more important role. Fleming was trained to read patterns and to question their underlying meaning (Williams, 1992). The circumstances around the discovery of penicillin, for example, were strikingly similar to his discovery of lysozyme (McFarlane, 1984). It is questionable whether the unusual appearance of the bacterial colony he cultured would have caught Fleming's attention if he had not compared it to the other colonies of bacteria that were free of the mold. It was an outlier.

In the same lecture, Fleming said that my "only merit is that I did not neglect the observation and that I pursued the subject as a bacteriologist. It seemed to demand my attention." He went on to say, "It is also probable that some bacteriologists have noticed similar changes to those I noticed, but, in the absence of any interest in naturally occurring antibacterial substances, the cultures have simply been discarded" (*Nobelprize.org*, 1945). Penicillin, had in fact been discovered earlier, but it is often easier as history has proven and Fleming confirmed, to neglect outliers. Fleming, however, like good bacteriologists, was trained to ask "Why?".

Who Would Have Lived Here?

If you hike through the New England woods, it is likely that you will encounter a stone wall. It will inspire you to ask, "Who would build a stone wall in the middle of the woods?" Like a good mystery, there are clues to who might have lived there.

The stone wall is cobbled together with large flat stones with smaller stones wedged in the crevices of the larger stones or laid on top. On one side of the stonewall, there are numerous white pines growing along with a few white birch trees. The land is relatively flat. As you climb over the wall to the other side, the land shows a different history. Instead of being flat, the land shows a succession of bumps and depressions, much like moguls one would find on a ski slope. The pattern of orientation is peculiar in that the succession of bumps and depressions all face in one direction. Scattered throughout are well-established stands of broad-leafed trees, such as oak and maple (Wessels, 1997).

As you wander about, you stumble upon a cellar hole, likely, the remnants of a home once lived in by the family who worked the land (Schlobaum, 2014). Like the wall, the foundation is constructed of large flat stones wedged into the surrounding soil. The foundation is about 5 feet deep by 35 feet wide. A large stone foundation sits in the center of the cellar hole. Just outside one of the walls is a refuse pile of metal, glass, and china. The widest part of the cellar holes that have been researched face in the same direction. (Schlobaum, 2014). Next to it are two very large sugar maple trees that seem to be found together at each cellar hole site. Our discovery inspires many questions including: Who lived there? Was it a farm? Which way did it face? When was it abandoned? Why was it abandoned?

To answer these questions is akin to piecing the parts of a puzzle together. Beginning with the original question, why would anyone build stone walls in the middle of a forest? The answer is that the forest was not there at the time (Wessels, 2008). Walls were built to keep farm animals either in or out. The stone wall separated the land for different purposes. During the 19th century, land was cleared to grow crops or

Figure 52: Who would build a stone wall in the middle of the woods? Like a good mystery, there are clues to who might have lived there. Image courtesy of Shutterstock.

provide pastureland for grazing animals, or was mowed to produce hay. To use the land for growing crops, rocks had to be removed from the soil. Instead of carrying them off, settlers would pile the larger ones together to form a stone wall.

The side of the wall where the land was relatively flat was likely used to cultivate crops. Evidence for this is finding smaller stones wedged in the crevices of the stone wall's larger stones or laid on top. When the soil is turned each year to cultivate a new round of crops, new stones appear. They have to be removed, so farmers added them to the stone wall, making it larger. The other side of the wall is a different story (Wessels, 1997).

Evident from the pattern of bumps (pillows) and depressions (cradles), the land could have been either used for forestry or left untouched. "The bumps and depressions are the graveyards of trees felled by wind. As a tree is blown over, its roots are ripped from the ground carrying a large amount of earth and stone attached to its roots. The removal of roots from the soil excavates a depression called a 'cradle.' Over time, the roots and trunk decay, dropping the excavated earth and stone next to the cradle, creating a mound or 'pillow' (Wessels, 1997). That the pillows and cradles were facing in one direction suggests that the trees were victims of a "blow down." Unlike tornadoes, blow downs are winds of high force blowing from one direction that topple trees in the same direction.

The settlers were early environmentalists. They practiced passive solar heating, for example, by always siting the widest part of the house toward the south. The south-facing side had the greatest number of windows to catch the sun's warmth in the winter; the north-facing side had the fewest. Two large sugar maple trees were often found growing on the south-facing sides to take advantage of the shade they cast in the summer. In the winter, the sun shone unobstructed through their bare limbs to heat the house (Schlobaum, 2014). Food that was left over was recycled as fertilizer or fed to farm animals. The size of the cellar hole was sometimes an indicator of how well off the family was. A cellar hole that was 35 feet wide was one of the largest one could expect to find. A stone foundation in the center or near one end of the cellar hole indicated a chimney. A double chimney indicated that the family was well off (Schlobaum, 2014).

Reading the landscape and reconstructing the past takes certain skills and talents. According to Tom Wessels, these skills include the ability to see changes in large-scale patterns, develop hypotheses to explain the changes, look for small-scale evidence to support or reject the hypotheses, and use common sense. "Most people can't see the forest for the trees in front of them. Most see the forest as a blur. It is pretty, but too complex to make any sense of it" (Wessels, 2008). In a way, *reading the landscape* is an apt term since it is like learning words so that one can read the paragraph. As people learn to identify different trees, discern different shapes of trees, see the way stone walls are built, observe whether the ground is smooth or pillowed and cradled, they start to see that the forest is not a blur, but that two different parts can have a different history. Once they can distinguish these differences, the homogeneity becomes replaced by an ever-changing series of forest patterns that is more complex, but at the same time understandable (Wessels, 2008).

For this story, "Who Would Have Lived Here," I interviewed Tom Wessels and Dietrich Schlobaum. Both are professor emeriti, Tom at Antioch College in New Hampshire and Dietrich at Springfield College in Massachusetts. It was interesting and fascinating to find that both shared similar characteristics. Both are nature smart; Tom (part geologist, naturalist, and anthropologist) focuses on large-scale forest patterns; Dietrich (part historian, part naturalist, and archeologist) focuses on smaller-scale patterns, such as part of cellar holes. Both are highly curious, nature smart, and highly motivated to learn what the patterns that they recognize tell them.

Nature Is Taking Notice

On my winter walks through a local park, I am always vigilant and excited for the signs of spring. One of the first plants to push through the soil, often snow covered, is a purplish green plant called the skunk cabbage (Holdrege, 2000). Sometimes, I smell its arrival before I see it. The pungent odor that smells like rotten meat attracts insects to its flower. It is one of those rare plants that is warm-blooded and able to heat its body above the ambient temperature. Its warmer body temperature not only enhances its attraction to insects,

Figure 53: Skunk cabbage after snowstorm. It is one of those few plants that is warm-blooded and able to heat its body above the ambient temperature. Its warmer body temperature not only enhances its attraction to insects, but also helps to melt its way through the frozen ground and snow. Image courtesy of Shutterstock.

but also helps to melt its way through the frozen ground and snow (The Nature Conservacy, n.d.).

Since I first logged its spring arrival in my journal some 25 years ago, skunk cabbage now emerges about five days earlier, near the middle of March. While few were noticing, these lowly wetland plants were slowly signaling that something out of the ordinary was happening to the climate. Nature was taking notice. It was getting warmer.

The skunk cabbage is not alone in getting a head start on spring. The first flowering date for dozens of the most common spring flower species in the northeast United States now occurs earlier than past years (Elwood, 2013). Since 2003, Dr. Richard Primack of Boston University has been studying the pattern of flowering of plants living around Walden Pond in Concord, Massachusetts. Walden Pond is the same site that Henry David Thoreau recorded notes of wildlife beginning in the 1850s. By studying the same plants that Thoreau did, Primack was able to construct the phenology of flowering of 43 of the most common plants spanning 161 years (Parry, 2012). Phenology is the study of periodic plant and animal life-cycle events and how they are influenced by seasonal variations in climate. During the century and a half since Thoreau first recorded first flowering in his journal, the temperature has warmed more than 4° F and flowers have bloomed on the average seven days earlier; some now are blooming a month earlier (Miller-Rushing, 2012).

The evidence for rapid climate change is compelling. The current warming trend is proceeding at a rate that is unprecedented in the past 1,300 years. All major global surface temperature reconstructions show that Earth has warmed since 1880. Most of this warming has occurred since the 1970s, with the 20 warmest years having occurred since 1980. It is even more alarming to learn that 10 of the warmest years occurred in the past 12 years (Jenkins, n.d.). The causes for climate change are controversial. The evidence that humans are responsible is so far circumstantial. However, scientists, acting like detectives are searching for fingerprints as evidence for human involvement. The evidence is growing that carbon dioxide and other greenhouse gases are responsible (Jenkins, n.d.; Kahn, 2013). Nature is taking notice. As the big picture of climate change comes more into focus, we are seeing and feeling the ripple effects. Some changes were expected, while others were surprises.

Peanuts in New England? At Drumlin Farm Wildlife Sanctuary in Lincoln, Massachusetts, staff had some fun trying to grow plants in New England that were typically found in the southern United States. They were able to grow cotton and peanuts, crops typically found much farther south. Others pointed out that they have been planting their own gar-

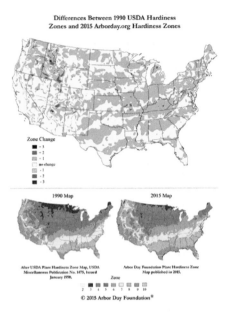

Figure 54: Plant hardiness zones 1990 and 2015. Plant hardiness zones are recommendations for planting based on the risk of extreme cold in a given region. Warmer zones (higher numbers), for plants less hardy to deep freezes, are typically found farther south. Zones (lower numbers) prone to harsh cold snaps are typically found farther north. Note that the hardiness zones have shifted further north over time. Plants suited to grow in the southern climates can now survive farther north (Brown, 2017). Image courtesy of USDA and Arbor Day Foundation.

Figure 55: One significant environmental change is an increase in airborne pollen count. Pollen from pine trees blanket a car in North Carolina. A note in the dust on the hood reveals thickness. Image courtesy of Shutterstock.

dens earlier and earlier over the years. Many noted that they start their tomatoes two weeks earlier than they once did (Brown, 2017).

Gardeners were some of the first to discover the impact of the recent warming pattern when the United States Department of Agriculture (USDA) published the 2015 plant hardiness zone map. Plant hardiness maps tell gardeners what plants are expected to thrive in the zone where they live. The USDA hardiness map divides North America into different latitudinal zones, each representing a 10ºF range of "average annual minimum temperature"—the coldest lows that can be expected in that area. The map now reflects how rising temperatures have shifted planting zones northward. Re-cataloging a gardener's yard into a warmer area opens new options for planting plants and shrubs that would probably not have survived local winters 30 or 40 years ago (Weeks, 2009; Weeks, n.d.).

Are you sneezing more often? The occurrence of allergic and asthma diseases is skyrocketing. When the search interest for the word "allergy" is plotted using Google Trend from 2004 to the present, the search interest almost doubles. Some estimates are that as many as one in five Americans now have an allergic condition. Allergies are unpleasant immune responses that are triggered by naturally occurring substances, such as pollens or other influences in our environment. Similarly, the global rates of asthma have increased significantly between the 1960s and now. It is recognized now as a major public health problem (National Wildlife Federation, n.d.).

Changes in genetic factors are unlikely to be the underlying cause of the rise in allergic diseases, since the increases in allergies and asthma occurred relatively rapidly. This suggests that something in the environment is responsible. One significant environmental change is an increase in airborne pollen count (Staudt, n.d.). A pollen count is the measurement of the number of grains of pollen in a cubic meter of air. Brace yourself if you suffer from allergies or asthma provoked by airborne pollen. New research suggests that allergies triggered by pollen are set to increase—in both duration and severity—with climate change (Koch, 2013). Pollen counts sampled in the northeastern United States averaged about 8,455 in 2000. They are expected to surpass 11,412 by 2020 and will top 18,285 by 2040—possibly pushing as high as 21,735. Not only are the average pollen counts likely to increase dramatically, but the allergy season is also set to start much earlier in the year and end later. In 2000, for example, pollen production began around April 15, but in 2020, it was expected to begin around March 27 (Courage, 2012).

Rising temperature and carbon dioxide levels is the one-two punch that experts think causes the increased pollen levels. Atmospheric carbon dioxide is now 40 percent higher than it was a century ago. Higher levels of CO_2 in the at-

Figure 56: Display of Optimum, Current, and Projected CO_2 Concentration in the Atmosphere. Image courtesy of Shutterstock.

mosphere act as a fertilizer for plant growth (Taub, 2010). A 2002 study in the Annals of Allergy, Asthma, and Immunology found that ragweed, which causes most fall allergies, produces 61 percent more pollen when grown in an atmosphere with double the normal amount of carbon dioxide. Warmer temperatures, together with higher carbon dioxide, cause most plants to grow faster, bloom earlier, and produce more pollen (Valentine, 2013).

Why are allergies worse after a rainstorm? For some people, right after it rains, throat, ears, and eyes itch almost unbearably. The symptoms go away the next day, as long as it does not rain again. For asthma sufferers, it is more serious. The association of thunderstorms with allergic reactions to pollen is a phenomenon coined "thunderclap asthma" (Hauser, 2015; D'Amato, 2006). The connection was first noticed when emergency room visits related to asthma attacks spiked after a rainstorm. This seems counterintuitive because a good rain helps clear pollen from the air. Nevertheless, it is also true that, in a rainstorm, pollen becomes saturated with water, expands and eventually fractures, releasing small particles into the air at very high concentrations. Then, when people inhale those particles, the pollen triggers allergic attacks (ExpressMED, 2013). So, while it is advised that allergy and asthma sufferers stay indoors on dry, windy days when the pollen count is typically high, experts also recommend staying inside after a rainstorm as well.

The rippling effects of climate warming will continue to be discovered in all likelihood. Rachel Carson reminded us that everything in nature is connected. This pattern of relationships between plants and the Earth, between plants and other plants, and between plants and humans creates a web of life that lives in a delicate balance. When even one of these connections is disturbed, life's web becomes unraveled, often leading to unexpected and surprising discoveries.

Choosing Green to Be Seen

When you say electric cars, what model comes first to mind? Is it Tesla? It's no surprise that most people think Tesla. Tesla was founded in 2003 by a group of engineers who wanted to prove that people didn't need to compromise to drive electric and that electric vehicles (EVs) can be better, quicker, and more fun to drive than gasoline cars. Tesla believes the faster the world stops relying on fossil fuels and moves towards a zero-emission future, the better. However, Tesla is not alone in building electric cars. EVs are becoming increasingly common, with many manufacturers currently offering models that plug in. Dozens more are expected to hit the market over the next few years.

What motivates many owners to buy EVs is often status. They want people to know they care about the environment (Maynard, 2007). The Tesla car is efficient not only because Tesla invested in material science, but also because they invested in behavioral science. The dashboard shows drivers how much energy they are saving in real time. "It makes former speed demons drive more like cautious grandmothers" (Alex Laskey at the 2020 TED Conference).

Status, in fact, is apparently at least or more important than money. New studies show that money and social values are processed in the same brain region, providing insight into how we make choices. The brain's striatum was previously recognized through magnetic resonance imaging (MRI) studies as the brain's monetary reward center. However, in situations where status replaces money as a reward, the brain's striatum also "lights up" when MRI peers into the brain. This led researchers to conclude that when we make decisions, status and money are weighed against each other (Swaminathan, 2008). Which one wins out?

This was answered in part across three studies of human behavior carried out by Vladas Griskevicius and his colleagues at the University of Minnesota (Griskevicius, 2010). Griskevicius studied the interplay between choosing pro social or pro self by asking people to choose between an environmentally beneficial product (designated a *green product*) or regular, more luxurious one (designated a *nongreen product*) when both were offered under three different hypothetical conditions. Researchers found consumers are willing to trade luxury and performance for the perceived social status that comes from buying a product with a reduced environmental impact.

In the first experiment, researchers designed an experiment to test if, given a choice, shoppers would favor green, environmentally friendly or nongreen luxurious products that were equally priced. For example, shoppers could choose between a deluxe dishwasher that had many bells and whistles versus an equally priced standard dishwasher that was designed to conserve on water and energy. Shoppers, the results show, chose to forgo the more luxurious nongreen product in favor of the green, environmentally friendly product (Griskevicius, 2010). In the second experiment, the conditions were the same as the first, except for one variable. Shoppers were given the opportunity to shop both in public and privately. When subjects shopped in public, the green environmentally friendly products were again preferred. However, when shoppers were given the choice to shop in private (such as online), they chose the more luxurious nongreen products (Griskevicius, 2010).

A third experiment showed that buyers with social status in mind preferred green products when they were more expensive than their conventional alternative. When the green products were priced higher than the nongreen products, they chose the green, environmentally friendly products. When the green products cost less, they favored the more deluxe products. Unless social status motives are considered, the results might seem contrary to what might be expected. If subjects chose the less expensive green products, they may be perceived as unable to buy the more expensive nongreen products. In situations where shoppers chose the pricier green products over the nongreen ones, they hoped to be perceived as altruistic (Griskevicius, 2010).

Discovery of the Genetic Stuff

The discovery of deoxyribonucleic acid (DNA) and its structure has had a groundbreaking impact on how guilt or innocence of the accused is decided, disease is diagnosed, and crops are made resistant to disease and pests. DNA contains the biochemical instructions that make each species unique. DNA, along with the instructions it contains, is passed from adult organisms to their offspring during reproduction, guaranteeing that elephants only give birth to little elephants, giraffes to giraffes, dogs to dogs, and so on for every type of living creature (National Human Genome Research Institute, 2012). DNA is the genetic stuff that all life, from simple yeast to humans, has in common.

Patterns run through the well-known discovery of the structure of DNA. Watson and Crick discovered the correct structure for the genetic stuff, inspired by piecing together several parts of the chemistry of the DNA puzzle already known. Thanks to the work of Rosalind Franklin, Watson and

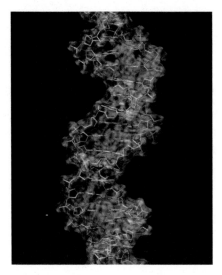

Figure 57: Patterns run through the well-known discovery of the structure of DNA. Equipped with information discovered about the three-dimensional structure and knowledge of the pattern of the pairing of chemical base pairing, Watson and Crick correctly identified the structure of DNA. Image courtesy of Wikimedia Commons.

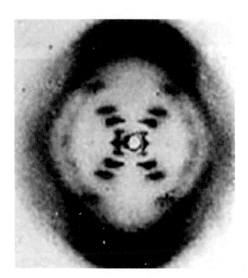

Figure 58: The most important photo ever taken. Thanks to the work of Rosalind Franklin, Watson and Crick first recognized the three-dimensional structure of DNA from this X-ray image. That image, distinguished by a peculiar pattern of dots, showed that DNA was a double-stranded helix. Image courtesy of College of London.

Crick first recognized the three-dimensional structure of DNA from an X-ray image.

That image, distinguished by a peculiar pattern of dots, showed that DNA was a double-stranded helix. The question of how the helix was held together was answered in part by the pattern of chemical bases that made up the DNA molecule. Equipped with information discovered earlier about the three-dimensional structure and knowledge of the pattern of chemical base pairing, Watson and Crick correctly identified the structure of DNA. For their efforts, Watson and Crick received the Nobel Prize in Physiology or Medicine in 1962.

Watson and Crick's groundbreaking discovery raised many questions about DNA that were left for others to answer. For example, how are the chromosomes that carry our genes copied in their entirety during cell division and protected from breakdown? Otherwise, the genes and cell would be damaged after several divisions. Elizabeth Blackburn, Carol Greider, and Jack Szostak answered that question, and received the Nobel Prize in Physiology or Medicine in 2009. The answer lay at the end of the chromosome, in a structure called a *telomere* and an enzyme that forms the ends called the *telomerase*.

Blackburn discovered that the DNA at the end of the chromosome had a unique sequence of chemical bases—CCCCAA—which was repeated over and over again. Szostak learned through his work that the unique sequence of bases gave the telomere its protective powers. He also discovered, that the *pattern of bases* was not just unique to single species, but commonly occurred from amoeba to humans.

And, Greider discovered the enzyme that was responsible both for the synthesis of the new ends of chromosomes and for the unique sequence of bases. On one Christmas morning,

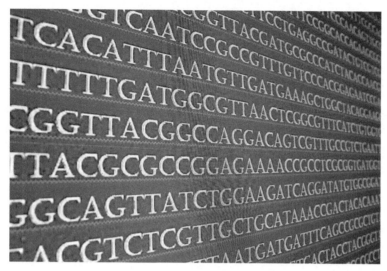

Figure 59: Pattern of DNA bases; A (adenine) pairs with T (thymine) and G (guanine) pairs with C (cytosine). Image courtesy of Shutterstock.

she observed the first proof of the enzyme telomerase. Using a technique to separate various chemicals from each other on a gel, she spotted a ladder-like pattern never seen before. It was the enzyme responsible for synthesizing the sequence of bases unique to the telomere. Today, there is excitement over the possible role the telomere and telomerase play in the aging process and cancer.

Smiley Face Rewards

My wife and I are conscientious about turning lights off in rooms that are unoccupied. It wasn't always like that. We changed our behavior when we received an energy audit from our electric company, comparing our energy use to that of our neighbors. The audit compared our electrical use to 100 nearby homes that are similar in size to ours. The audit revealed to us that we used 12 percent less energy than our most efficient neighbors and considerably less than that for all 100 neighbors. Our rank was 15 out of 100 of our neighbors with 1 being the most conservative. Best of all, we were rewarded with two smiley faces!

The company behind these reports is Opower, founded in 2007. Opower's data has shown that highlighting patterns in consumers' energy usage translates to measurable energy savings. Some of these data-driven patterns that are used on reports include:

- a single-period neighbor comparison,
- neighbor comparison from the previous winter,
- neighbor comparison from the previous summer,
- neighbor comparison over the past year,
- comparison to the customer's own energy usage from the previous year,
- a daily breakdown of a customer's energy usage by time, and
- breakdowns of home energy usage by appliance.

Opower helps customers save energy and money just by highlighting these energy-usage patterns and presenting tips for how customers can reduce their energy usage. Savings are consistent across all income levels (Bend, 2014).

I interviewed Dave Bend of Opower, who told me that the report was inspired by a social experiment done several years ago by two graduate students. The premise behind the social experiment was to explore how to get people to start paying more attention to the energy they were using, and how to motivate people to start wasting less energy. The behavioral science experiment was designed by graduate students in 2003. The graduate students put signs on every door in a neighborhood in San Marcos, California, asking people to turn off their air conditioning and turn on their fans (Bend, 2014).

They used door-hangers with a different messaging tactic for each of their four experimental groups.

The first technique was a money-saving message, the second was an environmental message, the third told people they should be good citizens and help prevent blackouts, and the fourth said, "When surveyed, 77 percent of your neighbors said that they turned off their air conditioning and turned on their fans. Please join them." (Bend, 2014).

The only messaging tactic that had any impact on energy consumption was the fourth message. The people who received this message showed a marked decrease in energy consumption simply by being told what their neighbors were doing. This idea that social pressure can motivate people to save energy prompted Alex Laskey and Dan Yates to found Opower, and it is still the ideology behind Opower's reports today. Opower can receive feedback from customers through the web portal, which allows customers to share their story about their experience with the program (Bend, 2014).

The psychology behind the reports somewhat taps into social pressure and competitiveness, but Opower's primary goal in providing the reports is to provide insights and context about a home's energy use so that customers can save money and energy. Customers are recognized for their energy-saving efforts by receiving positive feedback on their reports. For example, customers who save more than their efficient neighbors, which is the most efficient 20 percent of neighbors, are congratulated on their reports. Additionally, in order to highlight program success, Opower can spotlight a customer who has had a positive experience with the program and feature their testimonials in a press release or in the media.

The Opower platform consistently saves customers 1.5–3.5 percent on their energy bills. In aggregate, Opower has saved 4.2 terawatt hours, enough to take a city the size of Las Vegas off the grid. Opower has also saved utility customers $464 million and has abated 6.4 billion pounds of CO_2. It seems counterintuitive that an electric company that sells electricity would ask its customers to conserve energy. State regulators in many states set efficiency mandates that require utilities to reduce consumers' energy consumption by a certain percentage, usually around 1 percent of sales. Opower provides a way for utilities to meet these mandates by providing large-scale energy efficiency through behavior change (Bend, 2014).

In addition to helping utilities meet efficiency mandates, Opower's platform also increases customer satisfaction with the utility, which is becoming increasingly important to utilities as the utility industry faces a dramatic transformation. This transformation is expected to happen within the next 10 years, causing utilities to build closer relationships with their customers to meet the challenges in the market and remain viable. This means being transparent about energy usage and providing personalized advice to every customer. Opower is well positioned to help utilities meet these challenges by providing measurable results and meaningful customer experiences (Bend, 2014).

Word-Smart—Word Patterns That Inspire Discovery

"A man reached for a fly, fell backwards, and, unfortunately, died." How did the man die? This puzzle is an example of what some call a *minute mystery*. Each puzzle describes an unusual scenario, and your job is to figure out what is going on. The mysteries take only a minute to tell but one can take as long as needed to solve it. One can ask questions, but only questions that can be answered with a yes or a no answer. The key to cracking the minute mystery is contained in the pattern of words used to create the mystery.

You would not have been alone in guessing wrongly that the man was trying to catch a housefly. The word "fly" has more than one meaning. The fly can mean a housefly, a fly baseball, or the zipper on men's pants. The fly in this case is a fly baseball. A baseball fan had an unfortunate fall reaching backwards to catch a fly ball. My research shows that children as young as six years old can learn the pattern. After learning the pattern, my six-year-old granddaughter, Bryson, with a

Figure 60. Word patterns can detect trends in history and culture as well as reveal our inner thoughts and personalities. Wordle, for example, is an app for generating "word clouds" from text that the user provides. The clouds give greater prominence to words that appear frequently in the source by boldfacing them or enlarging them. This word pattern illustrates that words like science, college, and degree are closely associated with the word "education." Image courtesy of Shutterstock.

little coaching about words with double meaning invented the "fly" mystery.

Recognizing word patterns is important from learning how to spell and read to discovering the origin of languages. Questions from who wrote Shakespeare to who was the Unabomber can be answered by carefully studying word patterns. Word patterns can detect trends in history and culture as well as reveal our inner thoughts and personalities. Wordle, for example, is an app for generating "word clouds" from text that the user provides. The clouds give greater prominence to words that appear frequently in the source. This word pattern illustrates that words like science, college, and e-learning are closely associated with the word "education." These are just some of the things that can be discovered by recognizing and interpreting word patterns.

WORD PATTERN ACTIVITY

"What's The Deal With The Uke?"

ENGAGE. "What's the deal with the uke?" That's what my instructor asked me during my banjo lesson. It puzzled her why so many of her new students wanted to learn how to play the ukulele (uke). Moreover, her long-time banjo and guitar students were now showing interest in playing the uke. I was among them.

EXPLORE. This change in interest is documented worldwide through a word search with Google Trends (*www.google.com/trends*) beginning in 2004 and continuing until now .

EXPLAIN. The National Association of Music Merchants reported a 54 percent jump in ukulele sales in 2013 that can

Figure 61. The popularity of the ukulele and banjo over recent time. Image courtesy of Shutterstock.

be traced in large part to the instrument's accessibility, affordability, YouTube popularity, and celebrity esteem (Jacobson, 2015). Its four soft nylon strings, small body, and affordable cost invite novices to try it. It is far less intimidating than taking on the cumbersome electric guitar. It is often the instrument that primary grade students first learn to play music (Sidwell, 2013). George Harrison of the Beatles was a big promoter of the uke, saying, "Everybody should have and play a uke. It's so simple to carry with you, and it is one instrument you can't play and not laugh! It's so sweet and very old" (Jacobson, 2015).

EXTEND and EVALUATE. These forecast trends for both banjo and ukulele continue into the future. What effect will this have on manufacturers of both instruments? Will this motivate stringed instrument instructors to change their practices? What will music retailers do in view of these trends. Do you see a new genre of music in the future?

The following exercise asks questions to determine whether you are word smart. Check off the questions that you answer with "yes." If you answered "yes" to a majority of questions, you are especially good at recognizing word patterns.

Can you/Do you

- perceive what funny jokes have in common and create one?
- notice why many jokes use words with the letter "k"?
- recognize that languages are not static but, rather, evolve much like biological species?
- recognize that languages much like biological species have a common ancestor?
- recognize that the frequency of words we use uncovers our feelings, our self-concept, and our social intelligence?
- understand that our dialect words, slang, even the way we use punctuation are as distinctive as a fingerprint?
- learn from observing infants how they learn to segment human speech into separate words?

The Formula for Fun

Are computers funny? They can be. "Constructing a joke often means following a pattern and, as we know, computers are good at following patterns" (White, 2014). Scientists at Scotland's Aberdeen University for the Science Centre developed the Joking Computer.

The software was originally written for children with developmental disabilities, such as cerebral palsy, to help them develop language skills and offer original jokes to tell their

Figure 62. Our dialect words, slang, even the way we use punctuation are as distinctive as a fingerprint. Image courtesy of Shutterstock

Figure 63. Are computers funny? They can be. "Constructing a joke often means following a pattern and, as we know, computers are good at following patterns." Image courtesy of Shutterstock.

family and friends (Ritchie, 2011; White, 2014). Here is one created by the Joking Computer.

What kind of murderer has fiber? A cereal killer.

Here is a joke that I created using the software provided by the Joking Computer.

What do you call a rails joke? A trick track.

These are examples of puns, the most basic form of a joke (Wiseman, 2007). The Joking Computer builds jokes in a few stages using a very large dictionary and a small collection of rules. Stage 1 creates the answer parts; in my example above, I selected "track" from a random choice of words. In Stage 2, the computer asked me to select a word from a choice of words that sounded like the start of "track." I selected "trick." Next, for the question, I selected a word that described what kind of thing a "track" was. I chose "rails." In the last part, I selected a word from a group of words that meant roughly the same as "trick." I chose "joke." Finally, the stages were put together in predetermined pattern to create the "trick track" joke (Masthoff, 2011).

Computers can create jokes, but they have a much more difficult time identifying the humor in a joke. When scientific surveys compared them to human-created ones, human-created jokes were far funnier and won easily over computer-generated ones. The survey was carried out by LaughLab, created by Richard Wiseman to discover the funniest joke in the world (Wiseman, 2007; Noë, 2014).

LaughLab was created in 2001 with the scientific mission of finding the joke with the maximum mass appeal. Wiseman set up a website that had two sections. In one, he invited people to send in their favorite jokes; they were then posted on the LaughLab website (_laughlab.co.uk_). In the other section, people could rate the jokes on a scale from 1 to 5, five being the funniest. More than 40,000 jokes were submitted, and millions of people rated the jokes. Along with the ratings, Wiseman collected data on the evaluator's age, nationality, and gender. The joke that received the highest rating is this:

Two hunters are out in the woods when one of them collapses. He doesn't seem to be breathing and his eyes are glazed. The other guy whips out his phone and calls the emergency services. He gasps, "My friend is dead! What can I do?" The operator says "Calm down. I can help. First, let's make sure he's dead." There is a silence, then a shot is heard. Back on the phone, the person says "OK, now what?"

What makes this funny is the element of surprise: We are caught off guard when the hunter shoots his companion. What jokes, puns, and minute mysteries all have in common is the element of surprise. They lead our minds in one direction and, then, surprise you with the unexpected. "It's like magic, only with words" (Corley, n.d.).

Surprise and how hard we laugh is proportional to the degree of difference between what we expect to happen and what actually happens. I can relate to this through an experiment in which I took part. The experiment was to test the elasticity of several balls composed of different materials. The process was to drop the balls from a set distance between my hip and the floor and measure the height of the bounce. The first few balls proved to be quite elastic. They bounced at increasing heights, some reaching heights almost to the point from which they were dropped. The last ball landed with a thud, never leaving the floor. It so surprised me that I laughed out loud and so did everyone else who was observing the experiment. Goran Nerhardt explained why in a similar study he did with weights (Nerhardt, 1970).

Nerhardt wanted to know if he could make people laugh performing an experiment with balls of different weights. Subjects were asked to lift balls of varying weights from two to six pounds and judge which ones were lightest and heaviest. The final ball tested was much lighter than all the rest. Every subject laughed, even though the experiment was not designed to be funny. In fact, the greater the weight differed from what was expected, the more it caused subjects to laugh even harder (Nerhardt, 1970; Weems, 2014). Whether it is at weights, balls, or jokes, people laugh when the results differ from what they expect to happen.

Our brains are built to be pattern detectors. When the brain receives new information, it draws on prior experiences to interpret the new information. Most of the times the brain predicts the information correctly. When it does not, we laugh. Whether a joke or minute mystery, the formula for fun is the word pattern that creates a difference between what is expected and the unexpected. The greater the difference, the harder we laugh.

Laugh Lab's Richard Wiseman tried to better understand the formula for fun by answering questions like, "What makes

a joke funny?" Sometimes, just one word makes a difference. Here are two jokes that differ in just one word.

Two frogs were sitting in a pond. One said to the other, "rrrrivet." The other said, "I was going to say that."

"Two ducks were sitting next to a pond. One said to the other, "quack." The other said, "I was going to say that."

Do you agree with most people surveyed who claimed that ducks are funnier than frogs? We can only speculate why, but the k sound in both duck and quack seems to provide the answer. It may be due to something called "facial feedback." People feel happier by simply smiling. That is, people actually experience the emotion associated with their expression. Words with "k," such as *duck* and *quack*, cause people to raise the edges of the mouth into a smile. Consequently, smiling makes us feel happier (Wiseman, 2007).

Another question that Wiseman tried to answer was, "why is something funny to one person and not to another?" His research uncovered other patterns. Women laugh at jokes that make men look stupid. The elderly laugh at jokes that poke fun at their own infirmities, such as hearing loss and memory loss. Those who are powerless laugh at those in power. There is no single joke, according to Wiseman, that inspires everyone to laugh (Wiseman, 2007).

Just getting the right word pattern for a joke takes time. According to popular comedian Jerry Seinfeld, it can take a

Figure 64. What makes a joke funny? Sometimes just one word makes a difference. Image courtesy of Shutterstock.

long time. In an interview with *The New York Times*, Seinfeld described how he created a joke featuring his first encounter with Pop-Tarts (Seinfeld, 2012). "Normally, jokes take a few days to write, but the Pop-Tart joke was rewritten several times over two years" (Seinfeld, 2012). Seinfeld claims that to achieve the right sentence structure is much like writing lyrics for a song. Instead of matching words to a melody, words need to fit the timing and rhythm of the joke teller. The sentences cannot be too long or too short. Syllables of words are counted and letters are shaved off or added to words to achieve just the right timing and rhythm. If the joke is too long or too short by even a fraction of a second, it will fail to elicit laughter (Seinfeld, 2012). We now know that word patterns, whether created by a comedian or computer, are important to carefully consider when creating a joke or minute mystery.

It's a Miracle!

Babies come into the world ready to learn language, all 7,000 of them (Janson, 2012). They can potentially make and hear all the sounds in all the languages in the world (Werker, 1989). That is about 150 sounds in about 7,000 languages! Babies eventually learn to target the sounds that belong to the language they are hearing. In English, there are about 44 sounds (Bainbridge, 2012). Over time, babies lose the ability to identify the sounds of the other unspoken languages. By 12 months, infants learn how the sounds of the native language are put together to make meaning (i.e., words). In the months that follow, babies learn how to put words together to create sentences. Later, they learn by connecting the spoken word with the written word; they learn to read.

It's a miracle that children ever learn their native language. Infants enter the world hearing a jumble of new noises. Their

Figure 65. Babies need to decode noise into a language that permits them to communicate with their caregiver. Given that an infant enters the world hearing a jumble of new noises, it's a miracle that children ever learn their native language. Image courtesy of Shutterstock.

task is to decode the noises into a language that permit them to communicate with their caregiver.

We adults can appreciate the challenges infants have when we try to learn a foreign language. When I first tried to understand spoken Spanish, I perceived it as a fluid stream of unbroken sounds. I found it difficult, if not impossible, to perceive when a word began or ended. The pauses between the spoken words are missing, unlike when Spanish is written. It was only when I began building my vocabulary with more Spanish words that I could begin to segment the stream of sounds into separate words.

The challenge for infants to learn their native language is even greater. Unlike adults, infants do not have access to a glossary of words. Without a vocabulary to identify the words, how can infants distinguish when a word begins or ends? It is a basic chicken-and-egg problem: What comes first? Unless infants know the words, they cannot learn the language, and unless they know the language, they cannot distinguish the words. This dilemma is solved through pattern recognition. Babies' brains are built to be pattern recognizers.

In the English language, most words are two syllable words. Ninety percent of those words are stressed on the first syllable (Byrd, 2010). Words like *DOCtor* and *BAby* conform to the strong-weak pattern of English words. When babies encounter the first syllable of a word that is stressed, they get a "heads up" that a new word is beginning (Daland, 2009). Some English words like *guitar* and *languages*, however, stress the second syllable (guiTAR), and other languages show no preference for stressing either the first or the second syllable (Byrd, 2010; Werker, 2003). Other means for segmenting sentences into words, however, are at infants' disposal.

By nine months, infants learn that certain sounds go together. Take for example, the phrase, "baby hungry?" There are two sounds in the word *baby*; the first is *ba* and the second is *by*. Called phonemes (Werker, 1989), the sounds ba and by are among the smallest units of speech in the English language (Eimas, 1985). The two sounds are often near neighbors in the English language (i.e., they are usually found together inside a word). The word *hungry* consist of two other sound units, *hun* and *gry*. Likewise, the two phonemes are paired more than they are separate inside a word. After hearing the phase, "baby hungry," repeatedly, infants eventually learn which sounds are paired and which are not. Those sound patterns, which are not normally paired, inform infants of word boundaries, when a word is ending and another word is beginning. For example, the second syllable of *baby* (*by*) and the first syllable of *hungry* (*hun*) are rarely paired. When infant encounter the two sounds, *by* and *hun*, together, they recognize them as an outlier or an unfamiliar pairing. This word pattern becomes yet another clue to a word boundary (Saffran, 1996).

Because an infant's abilities to segment words improves when they hear words repeated, experts advise reading aloud often to infants. When I read a new book for the first time to my grandchildren, I try to speak as fluently and clearly as possible. Occasionally, I will stumble over a word or pause with an "uh" when I approach a new word. These pauses, filled with "uh" or "um," are called speech disfluencies and occur in highly predictable locations. For example, they can occur before unfamiliar or infrequent words, and before words that have not been previously mentioned in the discourse. Despite my fleeting embarrassment, recent research shows that it is all right to fumble a word occasionally (Kidd, 2011). The research even seems to recommend it. Caregivers, the data shows, can unwittingly cue infants that a new or unfamiliar word will follow and enhance infants' abilities to identify new words and help segment sentences into words.

The step from speaking words to reading words that connect symbols to sounds, sounds to words, words to meaning, meaning to memory, and memory to thoughtful information is no less a miracle (Willis, 2008). Reading is not a natural part of human development. Unlike spoken language, reading does not follow from listening, observation, and imitating others. There are no specific brain regions dedicated to reading comparable to those parts of the brain that process oral communications. Instead, reading draws upon multiple regions of the brain that must be coordinated to permit children to read (Willis, 2008). Given that reading is not a natural part of human development, teachers and caregivers have an enormous responsibility to teach children how to read. It is for now a combination of the art of teaching and the science of how the brain responds to stimuli that will guide educators in finding the best ways to prepare lessons and use strategies in such a way as to promote students' success as they learn to read effectively and joyfully (Willis, 2008).

Like learning to speak, reading draws on the ability of children to recognize patterns. Even the youngest readers can recognize patterns and use categories to process new information. They can learn to find patterns in letters and words and use this information to read groups of words (e.g., *sun*, *fun*, and *bun* all contain the '-un' letter pattern or family). Young readers can also categorize words by sounds (e.g., short/long vowel words, rhyming/non-rhyming words) or by meaning (e.g., words that mean the same thing or words that are opposites) (Reading Rockets, 2011). Many teaching techniques, according to Sarah Blodgett, can help children identify patterns by making them more noticeable. Printed instructional material uses color, style, size, or font to make target patterns stand out visually from surrounding letters or words. The use of boldface and underlines help identify the long vowels and syllables in the following sentence (Blodgett, 2018):

The **pro**nun**cia**tion of **na**tion is **diff**er**ent** from **na**tion**al**, as well as the words r**ea**l and **re**ality, s**i**gn and **sig**na**ture**.

This highlighting technique can emphasize a pattern as small as a single letter or as large as a group of sentences. In addition to locating patterns, it can show relationships between text elements. For example, the relationship between the vowel and the final "e" in the silent "e" pattern can be illustrated through typestyle or color as in *fate*, *same*, and *pale*.

The challenge that infants have learning to speak and read their native languages often continues into the adult stage when we try to learn a new language. Individuals substantially differ in their capacity to learn another language. Whereas some adults assimilate a new linguistic system with relative ease, most others struggle (Science Daily, 2013). What determines this variability? What predicts successful and fast acquisition of an additional language? Who are those who have the childlike ability to learn language? According to new research led by Ram Frost, the ability to learn a second language may depend, like infants, more on the ability to recognize patterns and less on linguistic skills (Frost, 2013).

In Frost's study, researchers measured how American students recognized the structure of words and sounds in Hebrew over two semesters (Frost, 2013). The students simultaneously were tested on their ability to spot patterns in visual stimuli. Participants watched a stream of complex shapes, which were shown one at a time. What the students did not know was that the shapes were organized into eight triplets. The order of the triplets was randomized, but each triplet always appeared in the same sequence. After viewing the stream, students were tested to see if they had picked up the pattern. The results showed a strong association between recognizing patterns in the shapes and learning another language (Frost, 2013).

Government agencies and the military have taken advantage of this new information to develop a new pattern recognition test to screen personnel for their aptitude to learn a new language. Even before someone learns the first word of a new language, the new High Level Language Aptitude Battery will predict from the beginning whether someone will be able to acquire the vocabulary, accent, and grammar to become a near-native speaker (Diep, 2013). Those who can learn patterns, remember sounds quickly, and associate new information are much more likely to be in the high attainment group. The test can predict with 70 percent accuracy who will be a high language attainer (Erard, 2014). Before the armed services invest money and time in training someone to learn a new language, they now can identify who will be good candidates and who will not.

Watch Your Words

Have you ever listened closely to the patterns of sounds that people make when they walk? It is possible to single out distinct differences among people. Some footsteps are a rapid succession of sharp sounds, while others are soft and sustained, made when people shuffle their feet when they walk. While the posture, the speed at which a person walks, and the walking style may sometimes change according to a person's mood, all of us do have a distinctive gait that can influence how others perceive us (Pandit, 2013). For example, I associate the people who have a brisk, forceful gait as people who do not like stopping anywhere once they are aware of their destination. These are the enterprising, type A go-getters in my view. People who have a spring in their step, and walk with their head held high and shoulders back, often seem to have a positive approach to life.

Words, like footsteps, are windows to who we are, where we're from, our secrets, emotions, health, and our personalities. Sigmund Freud is one of the first and best-known researchers who discovered how words we use give away what we are really thinking (Pincott, 2012). The "Freudian slip" supposedly explains some of our surprising word choices.

President George H. W. Bush on thinking about the years working as vice president to Ronald Reagan said, "We've had some triumphs, made some mistakes. We've had some sex... uh...setbacks (Pincott, 2012)." Freud would explain the slip as representing some repressed thought. He might invite Bush to take a seat on his couch to question his relationships, for example, or his feelings about extramarital sex. Slips are sometimes inevitable and are nothing more than slips. We normally make one or two errors for every 1,000 words we speak. However, sometimes these slips reveal, upon digging deeper, an unconscious thought, belief, wish, or motive (Pincott, 2012).

What do you call a carbonated beverage? Depending on where you are from, some say soda, some say pop. To most, a long sandwich containing cold cuts is called a sub, but in New Orleans you will be ordering a po' boy and in Boston, a grinder. When you hold a sale of things you no longer need, what do you call it? Midwesterners call it a garage sale or yard sale, for example, while Easterners call it a tag sale. These are popular figures of speech according to an online survey conducted by Harvard's Professor of Linguistics Bert R. Vaux (Vaux, 2002). According to Vaux, regional dialect research has a wide variety of applications in commerce, entertainment, and even in fighting crime. According to Vaux, mapping the 14 different North American dialect patterns could assist in predicting where a person grew up. Such predictions could

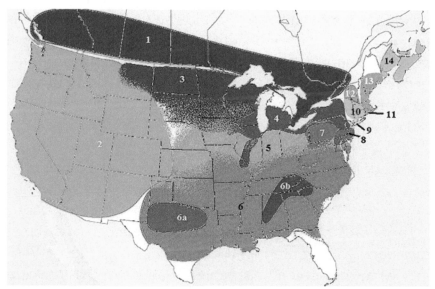

Figure 66. Fourteen different North American dialects. The study of regional dialect research has a wide variety of applications in commerce, entertainment, and even in fighting crime. Image courtesy of Wikimedia Commons.

be useful for law enforcement or marketing (Mohammed, 2003).

Not all words are created equal (Pennebaker, 2011). When we prepare what we want to say, we think first of the noun and verb we want to use. That is because the basic bones of a sentence are structured around a noun and verb. People focus on this type of heavy content words to conjure up an image of what, when, and where the sentence is about. When we then express what we are thinking into the spoken word, we add small words like articles, pronouns, and prepositions. These are the sentence's connective tissue, which link the bones of the sentence together. What makes them different from the content words is that they are added unconsciously. We have little mental control over them (Pennebaker, 2011).

The author of *The Secret Life of Pronouns*, James Penne-baker, calls these small words, like *the, a, I, me, you, we, of,* and *in,* "style" or "function" words. They make up just one-tenth of our vocabulary but are 60 percent of the words we use. Even though we rarely pay attention to them, style words uncover who we are and our emotions, social connections, and personalities. Take, for example, the prediction I made before the Super Bowl of 2019:

"I am confident that the New England Patriots will win the Super Bowl."

The words *confident, New England Patriots,* and *Super Bowl* identify the content of the sentence and inform the lis-tener what the sentence is generally about. Words that are branded as content words are nouns, verbs, adjectives, and adverbs. Pronouns, articles, and prepositions are the func-tion words; in other words, they are the words that make the sentence grammatically correct and manage how we express ourselves. In the sentence above, they are *I*, *am*, and *the*. Along with *I*, *am*, and *the*, they would include the words *you*, *a*, *but*, *for*, and *not*. If content words inform listeners what the sentence is about, function words shape how the sentence is communicated. If "the Super Bowl" were changed to "a Super Bowl," it would change the entire meaning of the sentence. Instead of implying that I am confident that the Patriots will win the upcoming Super Bowl, it implies that the Patriots are Super Bowl winless. Nevertheless, I am confident that they will win one in the future.

Language is the most common and reliable way for people to translate their internal thoughts and emotions into a form that others can understand. Words and language, then, are the very stuff of psychology and communication.

James Pennebaker, 2010

Pennebaker's scientific interest in what words say about us led him early in his career to study content words. The research led nowhere until he noticed a relationship between the use of small function words and personality types. Much to his surprise, he saw a recurring word pattern between the use of small words and someone's psychological state. His lifelong research of this word pattern reveals that small words speak loudly about our gender, emotional state, personality, age, and social class (Pennebaker, 2011).

To help analyze a wide variety of written material ranging from tweets and emails to books and poems, Pennebaker created a computer program called Linguistic Inquiry and Word Count (LIWC). The computer could count the number of words in any document and sort out the kinds of words used into both content and function words. It could identify the kinds of pronouns, articles, and prepositions. These small words became the clues to identifying difference between genders, personality types, age, social status, and emotional state.

One of the first significant differences he discovered between genders was their use of articles. Because men tend to talk more about things than women do, men use more articles. Guys talk about *the* ball game, *a* car, *an* athlete. Because they talk more about people, women use social words such as *them*, *him*, and *her* at far higher rates than men do. Men and women both use *we* at equal rates, but for different reasons. *We* can have two interpretations. When you hear, "we are in this together," it connotes that that the group is close knit and dedicated to working together. If, instead, someone says, "we need to work better together," it deflects responsibility from the speaker to the listener. Women tend to use the former meaning of *we*, and men use the "you, not me" meaning of *we* (Pennebaker, 2011).

Professor Pennebaker discovered other word patterns as well. Age brings on changes. Women as they age talk more like men: they use fewer social pronouns. Moreover, men talk even more like men. They use fewer personal pronouns in their conversations and more articles. Social class also makes a difference. People who feel subordinate to more powerful people pay more attention to themselves, evident from the use of first-person singular pronouns like "I and me" that they use in conversations with more powerful people. When he analyzed military transcripts, Pennebaker could successfully predict from their speech pattern the relative rank of two military members in conversation.

Have you found that perfect match yet? Speed dating is a type of matchmaking service that is popular among young men and women who are seeking lifelong mates. It works like this. Men and women are invited to meet each other over a series of short *dates*, usually lasting from three to eight minutes. After each interval, they then rotate to the next date. At the end of the event, participants submit their contact information to those who they would like to date in the future. Pennebaker recorded the conversations between people on speed dates. He fed these conversations into his LIWC program along with information about how the people themselves perceived the dates (Pennebaker, 2011). What he found surprised him (Spiegel, 2014). By comparing the small words used by two people in conversation, he could predict who would go on a future date. Men and women were more likely to date in the future when they used pronouns, prepositions, and articles at similar rates and in the same way.

LIWC may provide a unique insight into other written works. Such was the case with Pennebaker's study of al-Qaeda communications (Donges, 2009). In 2007 he and several co-workers, under contract with the FBI, analyzed 58 texts by Osama bin Laden and Ayman al-Zawahiri, bin Laden's second in command. The comparison showed how much pronouns are able to reveal. For example, between 2004 and 2006, the frequency with which al-Zawahiri used the word "I" tripled, whereas it remained constant in bin Laden's writings. "Normally, higher rates of 'I' words correspond with feelings of insecurity, threat, and defensiveness. Closer inspection of his 'I' use in context tends to confirm this," Pennebaker says (Donges, 2009).

Words not only reveal who we are as a person, but also provide the big picture of who we are as a society and culture of people. Google gave us the tools to view this big picture when they digitized more than five million books published between 1500 and 2008 (Karch, n.d.). Using key word searches, it is possible to determine the frequency of word or word-pattern use over time. David Brooks of *The New York Times* discovered a word pattern that has changed over time (Brooks, 2013). Words and phrases like "individualism, I can do it myself, and I come first" have increased over time at the expense of words like "community, tribe, and common good." The story that Brooks tells is that over the past half century, we have become a more individualistic society.

"As society has become more individualistic, it has also become less morally aware, because social and moral fabrics are inextricably linked. The atomization and demoralization of society have led to certain forms of social breakdown, which government has tried to address, sometimes successfully and often impotently (Brooks, 2013)."

My interest in sports led me to research the popularity of soccer, golf, and baseball over the past 50–60 years. I plugged the words, soccer, baseball, and golf into Google's Ngram program (*https://books.google.com/ngrams*), and I was not surprised by the results. A Google Ngram is a collection of written texts, especially the entire works of a particular author or a body writing on a particular topic. Since 1960, the frequency at which the word soccer appears in the literature has increased at the expenses of baseball and golf, which have declined in frequency. The number of golf courses that have closed since 1960 (Koba, 2014) and the decline of attendance at baseball parks confirm Google's word survey (Lindholm, 2014). World Cup soccer popularity on the other hand is at an all-time high in the United States. People are watching the World Cup in record numbers. The U.S. team's final match against Belgium in the 2014 World Cup had an overnight rating of 9.6 (16.0 million viewers) on ESPN alone, the largest rating ever for a soccer match on ESPN (Gaines, 2014).

Because they are like fingerprints that can reveal much about who we are, it's important to watch the words we use in our communications, such as conversation, blogging, and emailing. Words you use for Facebook (Cain, 2013), blogging (Boulder, 2011), even your resume (Pierce, 2013) can be mined for clues. They can be used to judge your age, sex, status, and relationship to other people, place of origin, even, some of your darkest thoughts. Working to be more sensitive of not only what you say, but also how you say it can improve your relationships with other people. If you are still looking for one, it can even help you find the perfect match to a lifelong partner. If you are interested to self-discover the word patterns that characterizes who you are or how compatible your language is to your partner's, use the websites cited below. If you are more interested in what word patterns say about a society over time, you will find directions at the Google's website that follows.

Use this website link to learn more about how to self-analyze your own word patterns using Language Inquiry Word Count (LIWC) *http://liwc.wpengine.com*. Use this site to determine how closely matched you and your partner are: Language Style Matching *http://secretlifeofpronouns.com/exercise/synch*.

Use this website to compare what word patterns say about society over time: Google's Ngram *https://books.google.com/ngrams*.

Words Patterns That Stir Emotion and Rouse Reason

What words stir emotion and rouse reason? The people most interested in knowing the answers to this question are marketers and advertisers whose job it is to search and find the "buy button" in customers and clients (Wells, 2003). The buy button is, unsurprisingly, seated in the brain. There are actually two buy buttons. One is located in the lower and the evolutionarily oldest part of the brain, in which a brain structure called the limbic system dominates (Brain Works Project, 2015). Composed in part by the thalamus and amygdale, the limbic system is responsible for our emotional lives and explains why things seem pleasurable or, alternatively, stressful (Lewis, 2013). The limbic system has been conceptualized as the "feeling and reacting brain." It explains why we sometimes are guided by our emotions rather than reason (Reed, 2007).

For example, my love of books sometimes leads me to buy books impulsively. It is the, "I need it right now!" syndrome. The title of the book might hook me, but, later, after opening the cover and reading the book, I may ask myself why I ever bought it. Moreover, I have bought clothes that were not even my size only because the sign said "on sale." The on-sale sign often causes an adrenalin rush and stirs the emotions of shoppers like me (Lewis, 2013). Studies have even shown that physiological changes in the heart, lungs, and skin happen when shoppers encounter this word pattern (Waldman, 2012).

Higher animals, including humans, fortunately also have a "thinking brain" which is more recently evolved (Reed, 2007). Its outward layer, called the *cerebral cortex*, covers the lower brain with a layer of neural cells. It is where our higher brain centers are located and where higher reasoning takes place (Lewis, 2013). When, for example, I buy a television, I do considerable research about its features, weighing the pros and cons. This kind of reasoning happens in the higher brain center. Advertisements that claim the television is a *Consumer Reports* best buy and has the best in industry warranty are word patterns that push my buy button. Both the higher and lower brain centers can operate independently or together. If, for example, I was pulled back from the brink of buying

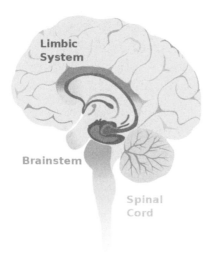

Figure 67. The Limbic System of brain is the evolutionary oldest part of the brain. Composed in part by the thalamus and amygdale, the limbic system is responsible for our emotional lives and explains why things seem pleasurable or, alternatively, stressful. Image courtesy of Wikimedia Commons.

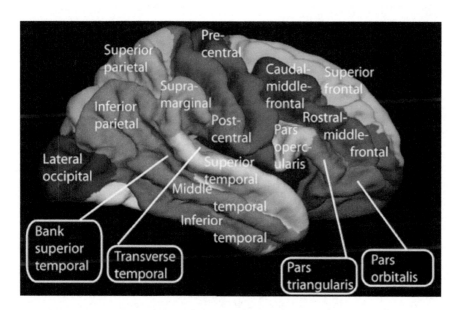

Figure 68. The Cerebral Cortex, the Thinking Brain. It is where our higher brain centers are located and where higher reasoning takes place. Image courtesy of Wikimedia Commons.

another useless book or a poorly fitted pair of pants by recon-sidering the purchases more carefully, it illustrates how the higher brain center can trump decisions made by the lower one.

A fascinating example of how the higher and lower brain centers interact to make decisions is the well-known rivalry between Pepsi vs. Coke. The rivalry received a lot of attention when Pepsi challenged Coke to a taste test in the 1970s. In a blind taste test, subjects overwhelmingly preferred Pepsi to Coke. Despite losing the taste challenge, Coke outsold and still outsells Pepsi two to one (Wells, 2003). What accounts for this paradox? The answer had to await a way to peer into the brain and observe it in action. This new way was a new technology called Magnetic Resonance Imaging (MRI). The technology uses a magnetic field and pulses of radio wave energy to make pictures of organs and structures inside the body. When the living brain is imaged under an MRI, the re-

Figure 69. In a blind taste test, subjects overwhelmingly preferred Pepsi to Coke. Image courtesy of Wikimedia Commons.

gions that are active light up due to blood flowing to those regions (Frank, 2009).

Curious about the Pepsi-Coke paradox, Read Montague, then a neuroscientist at Baylor College of Medicine, decided to repeat the taste test, but scan the brains of subjects simultaneously under the MRI. In a blind taste test, Pepsi was once again the preferred cola. When Montague examined the MRI images of the brain, the activity of the lower brain's limbic system was five times greater for Pepsi testers than Coke ones. The increase in limbic activity seemed to explain what underlay the emotional pleasure that subjects experienced from sipping Pepsi (Frank, 2009).

A remarkable thing happened, however, when Montague repeated the taste tests, but first showed the subjects what they were drinking. As before, subjects who were clued in to what they were drinking said they preferred Coke to Pepsi. The brain activity, however, differed greatly from the blind taste tests. Coke was preferred over Pepsi. The higher brain centers responsible for thinking and judging fired up, making that region of the brain's cortex light up in the scan at the expense of the lower brain centers. It shows that taste alone does not determine what consumers buy. The circumstances that form the setting for an event play a significant role in consumer tastes (Thompson, 2003). I think we already knew that. Doesn't the food you could have prepared at home, for example, always seem to taste better when enjoyed in the ambience of a nice restaurant?

What determines the loyalty of Coke fans that resonates with cola drinkers of all ages? What is it about the Coke brand that overrides our taste buds' yearning for sweetness? Brand loyalty is shaped by a combination of word, sound, and visual patterns. Coke's curvy script logo has remained almost unchanged since founding in 1886 (Stengel, 2013). The red color of their cans communicates love, warmth, and happiness. When Coke ads picture Norman Rockwell-esque paintings of Santa Claus drinking a Coke (Coca-Cola, 2012), it communicates, "Coke brings joy" (Volkmer, 2013).

The word patterns that light up the higher brain centers when people drink Coke are associated with images of optimism. Word patterns like "fun, simple moments of pleasure, authenticity, coming together, and uplift" that Coke has used to advertise its products signal hope for the future and a good time. Coke ads through words, sounds, and images inspire memories and other impressions of the drink—in a word, its brand—to shape preference (Volkmer, 2013).

If we know where the buy buttons are located in the brain, then, where could the "donate" button be located? Nonprofit organizations from hospitals to YMCAs to Habitat for Humanity want to know. Their livelihood depends on it. Their challenge to raise funds is more difficult than those that profit-making enterprises face. Because there is nothing concrete to put their hands around, donors typically receive little instant gratification (Entrepreneur, 2006). Because the brain's donate button likely targets both emotions and reason, the lower and higher brain centers will be involved. Word patterns created to target these centers can play a significant part in opening the wallets of donors. Here is one example of an appeal made by a hospital located somewhere in the U.S. Southwest.

Imagine the emergency room is crowded.

Your wait can be long.

The demand for surgery is growing.

There are not enough beds.

Imagine the emergency room is improved.

Your wait is shorter.

Surgical demand is met.

And your bed is waiting

These improvements happened because of your donation.

A pattern of three words when used consciously, correctly, and repeatedly can dramatically improve one's persuasive power (Tennant, 2012). What are the three words used in this hospital's campaign to raise funds?

The first is *imagine*. Instead of first asking for a donation, it is less challenging to ask a potential donor to imagine how upgrading the hospital's facilities will improve patient care. The soft-sell approach persuades donors to not only picture the improvements, but nudges the donor to feel a sense of ownership in upgrading the facilities. The second is *you*. *Your*, a substitute for *you*, is repeated several times. *You* acts as a placeholder for the name of the donor to insert. Dale Carnegie once said, "A person's name is, to that person, the sweetest and most important sound in any language" (David, 2015). It illustrates the benefits that a donor would potentially receive if she/he were a patient. The third word is *because*. From the time we become aware of the world around us, we are asking "why" questions. Learning why satisfies our need for order and a donor's need to know how a donation would improve hospital facilities and operations (Tennant, 2012).

We should never underestimate the power of words. A few properly chosen words can change history. President John F. Kennedy inspired thousands to volunteer for the Peace Corps by saying "Ask not what your country can do for you — ask what you can do for your country." These words have been repeated several different ways, but the message remains the same. Martin Luther King's "I Have a Dream" helped launch the Civil Rights movement. The often-repeated song "We Shall Overcome" has rallied people of color and many more to fight for justice over the years. More recently "hands up" has been the rallying cry in cities across the United States to call attention to racial injustice. When standing at the foot of the Berlin wall, President Ronald Reagan demanded Russian head of state Mikhail Gorbachev, "To open this gate, to tear down this wall." It sparked the collapse of the Communist governments of Eastern Europe and, eventually, the Soviet Union itself.

Picture Smart—Visual Patterns That Inspire Discovery

My fascination with letting images repeat and repeat—or in film's case "run on"—manifests my belief that we spend much of our lives seeing without observing.

Andy Warhol

When we were getting ready to sell our house, we hired Laura to stage our house for selling. Staging is a production, and the staging expert is the director. The house is being staged to look like a model home: Cozy, comfortable, colorful and inviting, with a personalized look to make it stand out from the rest of the other houses on the market. Staging can increase the value of the house and contribute to it selling more quickly. After a quick walk around our house, Laura generated a list of things to do. She explained that she was able, from the patterns of homes that had successfully sold, to picture in her mind what potential buyers hope to see in our own. We followed her instructions and, before we even put our home on the market, we had a purchase offer!

Our visual sense is highly tuned to spot patterns and can be trained to even greater levels of acuity that help us make sense of the world around us. When we look at a painting by the French painter Georges Seurat, we do not see a random assortment of dots, but an image depicting a casual afternoon in a Paris park or a riverbank scene (Artble, 2016). While we all have the talent to recognize visual patterns, it is enhanced in some people like Laura. Among others whose visual acuity is enhanced to recognize patterns are air traffic controllers. Air traffic controllers, who coordinate the arrival and departure of airplanes, issue landing and takeoff instructions to pilots, monitor and direct the movement of aircraft using radar

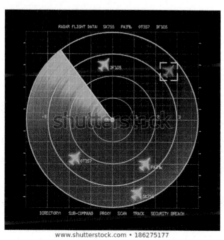

www.shutterstock.com • 186275177

Figure 70. Air traffic controllers, who coordinate the arrival and departure of airplanes, must perceive the flight paths of one or more airplanes, decide the best path for the planes to follow, and predict where the flight paths will take the aircraft. Image courtesy of Shutterstock

equipment, authorize flight path changes, and rely on pattern recognition. They must perceive the flight paths of one or more airplanes, decide the best path for the planes to follow, and predict where the flight paths will take the aircraft. Given their huge responsibility to control air traffic, it is no wonder that candidates to become controllers are screened for their ability to recognize visual patterns (Cook, 1993).

Can you read this?:

I cnduo't bvleiee taht I culod aulaclty uesdtannrd waht I was rdnaieg. Unisg the icndeblire pweor of the hmuan mnid, aocdcrnig to rseecrah at Cmabrigde Uinervtisy, it dseno't mttaer in waht oderr the lterets in a wrod are, the olny irpoamtnt tihng is taht the frsit and lsat ltteer be in the rhgit pclae. The rset can be a taotl mses and you can sitll raed it whoutit a pboerlm. Tihs is bucseae the huamn mnid deos not raed ervey ltteer by istlef, but the wrod as a wlohe.

McCarthy, 2008

Most people can read the above paragraph. It demon-strates that, as long as the first and last letters of any word are correct, it does not matter in what order you place the rest. The word will almost always be understood. Reading experts discovered that the brain has the ability to fill in meaning through visual pattern recognition, using the acquired skills of context and syntax (Crace, 2003). It is like listening to music. We often need to hear only a few notes to recognize the song. Countless musical works are composed using only the basic seven notes of an octave, yet the pattern created by these simple building blocks distinguishes most melodies from one another.

It is the goal of marketers to motivate consumers to recognize how the parts of a brand's story connect to the whole brand similar to how parts of a melody jog our memory of an entire song and single letters remind us of an entire word. Instead of repeating a brand over and over to gain product recognition, marketers have recently made the most of pattern recognition to brand their products (Shillum, 2011). Creating a believable and consistent brand begins with the creation of coherent patterns. By using patterns, the brand is placed in something, rather than just on it. Instead of adhering to a single, centralized big idea, a brand must create coherence around multiple, smaller ideas. Patterns connect a brand's visual identity to its behaviors, its interactions to language, its global ideas to local actions, and its small ideas to one another (Shillum, 2011).

Take Amazon, for example. My youngest daughter plans birthdays for her children around themes. The theme for my five-year-old grandson's birthday was snakes. I told my daughter that I could help. When I was in the Amazon River basin doing environmental work, someone gave me the skin of an anaconda snake, one of the world's largest. I sent it to my daughter to use as a centerpiece to the party table. When the party was over, she explained to my grandson that they needed to return it. He became upset and demanded that she buy another snakeskin from Amazon. Even at a young age, the loyalty to the Amazon brand had already been established.

Jeff Bezos, the founder of Amazon, believes that "Our brand is what customers say about us after we leave the room" (The Juniper Company, 2013). Amazon's brand is all about customer care. When I see a UPS truck pull up in front of our house, I automatically think it is a package delivered from Amazon. From the insight that customers felt looted by paying delivery charges, Amazon came up with Amazon Prime.

Figure 71. Amazon smile logo. When I now see the Amazon smiling logo on a shipping box, it unconsciously connects to Amazon's pattern of care for its customers. Image courtesy of Shutterstock.

For an annual subscription price of $139, customers get one year of free two-day delivery. When Amazon discovered that customers doubted using e-commerce to buy items virtually, Amazon launched customer reviews. When customers see other customer reviews, they feel more confident in buying online. Amazon learned that customers shop online because it is easy and quick. Based on this insight, Amazon developed one-click ordering (Narayan, 2011). When I now see the Amazon smiling logo on a shipping box, it unconsciously connects to Amazon's pattern of care for its customers.

Discoveries inspired by visual patterns range from realizing how diseases are spread to learning how kids' drawings mirror their thinking. Snapshots of patterns of consumer spending taken over a century reveal how our spending habits have changed over time. New visual tools can now make the once invisible visible. New techniques, for example, can now create "see through" mice that will give scientists a new understanding of the three-dimensional structure of organs and how certain genes are expressed in various tissues. Images taken from space that show patterns of the living and nonliving world will turn your head. Some of us do not need special tools to make the invisible visible. Autistic people have a special talent for seeing patterns that are invisible or overlooked by everyone else.

VISUAL PATTERNS ACTIVITY

Serve and Volley

ENGAGE. Tennis lovers always look forward each year to the Wimbledon tennis tournament. Recognized as the oldest and most prestigious tennis tournament in the world, it is the only major tournament played on grass. Unlike other surfaces which are either hard or clay, grass shows a wear pattern from players running and sliding to reach the ball. Compare the photos taken during Wimbledon of the two courts. The one

Figure 72. The famous Wimbledon court where the world's best tennis players compete. The photo on top was taken during a recent match. The photo on the bottom was taken some time in the 1960s. Images courtesy of Tennis Hall of Fame, Newport, Rhode Island, and Shutterstock.

on the top was taken during a recent match; the one on the bottom was taken several years ago. It is curious that when the wear pattern is compared between the earlier years of tennis played on center court to recent play, there is a marked difference. What could account for this difference?

EXPLORE AND EXPLAIN. Today, the wear pattern is visible largely around the baseline (top photo). In the 1960s and 1970s, the wear was visible both at the baseline and around the net (bottom photo). We can conclude that while yesterday's players altered between coming to the net to volley and playing from the baseline, today's players spend most of their efforts returning ground strokes from the baseline. What caused this shift in play strategy? Is it due to changes in equipment or changes in players' skill and power or something else?

EXTEND and EVALUATE. Would a great touch player of an earlier era like Ken Rosewall find it difficult to survive nowadays if they were limited to explosive exchanges from the baseline? Because so few differences separate top players in technical and physical ability, will the future focus for competitive advantage be on sharpening mental skills? Will there be a rise in demand for sports psychologists to help players outthink opponents (McElhinney, 2013)?

The following exercise questions whether you are picture smart. Check off the questions that you answer "yes." If you answered "yes" to a majority of questions, you are especially good at recognizing visual patterns.

Can you/Do you

- See the big picture when only given the parts?
- Visualize an object in three dimensions?

- Think in pictures, rather than words or numbers?
- Place yourself in the shoes of another person or even an animal to perceive their viewpoint?
- Predict if a system will work or not by picturing it in your brain?
- Visualize Earth's position in space to explain the seasons?
- Predict from a four-year-old's refrigerator drawings how intelligent the adult will be?
- Understand how a child's drawing mirrors their thinking?
- Visualize what an empty room might look like when it is furnished?
- Excel at games like Pictionary?
- Realize by mapping outbreaks how a disease is spread?
- Give directions using cardinal coordinates (i.e., N, S, E, W)?

The Autistic Brain

Stephen Wiltshire had only 15 minutes to memorize seven square miles of the city of London from the heights of a helicopter. Then, he was given only five days to create a panorama of the city on canvas. The results were remarkable. The landscape and the buildings were accurately drawn, down to the exact number of building windows (The Human Camera, 2013). Almost every major building in the city was drawn to scale—from the Swiss Re Tower (better known as the Gherkin) to the high rises of Canary Wharf—with the number of floors and architectural features all recaptured in precise detail (Wansell, 2008). This talent earned Wiltshire the nickname "the human camera."

His talent for visualizing a landscape or building and recreating it from memory grew out of his experience of autism. Wiltshire as a child could not make eye contact with others, even his mother, nor could he interact socially with other children and adults. Typical of individuals with autism, he was highly animated, experienced emotional outbursts, and was slow to develop language skills. What unlocked him from his autistic world was his ability to draw. Cityscapes and buildings quickly became Wiltshire's artistic focus, possibly because they represent the kind of stability, solidity, and repetition that autistic people often crave (The Human Camera, 2013). Wiltshire now travels the world as a successful artist. In 2006, Wiltshire was appointed a Member of the Order of the British Empire (MBE) for services to art. In September 2006, he opened his permanent gallery in the Royal Opera Arcade, Pall Mall, London (Stephen Wiltshire, n.d.). You can visit his website at *https://www.stephenwiltshire.co.uk/originals-prints*.

Is there a link between autism and exceptional ability? Apparently, there is, according to Professor Michael Fitzgerald, who authored a book about it (Fitzgerald, 2004). Stephen Wiltshire joins others like Albert Einstein, Andy Warhol, W.B. Yeats, Lewis Carroll, and Charles Darwin who fit the profile of high-functioning autistic individuals or those with Asperger's syndrome. They are all portrayed as having had highly visual brains and were able to think in pictures. When non-autistic people look at an image, their brains are activated in regions that process both visual and language information. When autistic people look at the same image, the visual part of the brain is more active than the language one (Mottron, 2011). Research now shows that autistic individuals outperform typical children and adults in a wide range of perception tasks, such as spotting a pattern in a distracting environment (Mottron, 2011).

Einstein, for example, visualized himself riding on a beam of light, flying at the speed of light to see in his mind's eye what was happening in the universe around him. This thought experiment was inspired by a compass he received as a child.

It intrigued young Einstein that no matter which way he turned the compass, the needle returned to the same location (Einstein Thought Experiments, 2009). That observation changed Einstein's perception of how the world worked. He thought that to make something move, something else had to touch it. Einstein imagined that there must be some invisible force responsible for moving the needle. It launched a life's journey to find out what it was. The invisible force he discovered was radiation that makes up visible light and other radiation types of shorter and longer wavelengths. Einstein's classic thought experiment helped him to develop the Theory of Relativity. Saddling up to a light beam was something that he could never do in reality, but by imagining it, his creative thinking was stimulated to the insights and understandings of how light and time functioned. From those imaginings came his world-famous theories in quantum physics (Kapeleris, 2010). Einstein's ideas helped build spaceships and satellites that traveled to the Moon and beyond. His vision of the universe helped us understand the universe as no one ever had before (Berne, 2013). Like many autistic children, Albert had a major problem in playing with and relating to other children. He described himself as a real loner who felt like a stranger everywhere; he often threw temper tantrums. His memory for words was poor, reinforced by the fact that his speech development was delayed. Despite these challenges, he was able rise above them and make remarkable achievements in physics (Fitzgerald and O'Brien, 2007).

"Temple Grandin is a rock star" (*https://source.colostate. edu/temple-grandin*). No matter what country she is in, people seek autographs or ask to take selfies with her. Grandin is among Time magazine's list of 100 people who have had an enduring impact around the globe. She is the featured subject in the HBO documentary film Temple Grandin. Who is Temple Grandin? Temple Grandin is internationally recognized as a livestock equipment designer who had drafted elaborate drawings of steel and concrete livestock stockyards and equipment. Half the cattle in the United States and Canada are handled in equipment she has designed for meat plants. This is quite an accomplishment for someone who is autistic. Autism is a very diverse disorder ranging from someone who remains nonverbal with a very severe handicap to someone who displays mild autism. "She's one of those rare people with autism who can explain autism," says Lesley Stahl, who interviewed Grandin while reporting her recent 60 Minutes story "Apps for Autism." "She's a sort of interpreter of autism for the rest of us" (Stahl, 2012).

Temple Grandin thinks in pictures. It is no surprise, then, that the visual part of her brain is larger than normal and lights up when she recognizes visual patterns (Grandin and Panek, 2014). She can translate the engineering symbols on drawings into a remarkable picture of the finished structure. When the drawing is finished, Grandin can "play the video" in her head and "test" the equipment to see if the equipment will work (Grandin, T., n.d.). She solves problems non-sequentially. That is, Grandin puts the pieces of a puzzle together with little regard to the final image. When enough of the parts of the whole have been put together, she can visualize the final image. Pictures replace words. It's no surprise that nouns are easiest to visualize. When she graduated from high school and college, she visualized changes in the life that awaited her as passing through physical doors.

Draw a House-Tree-Person

Drawings tell a lot about what we are thinking. The patterns we draw on paper can lead to the discovery of our biases, feelings, relationships, family, home life, self-esteem, and even intelligence. They are mirrors to our minds (Buck, 1984).

For some pediatricians, it is standard practice to ask children to draw. When children draw, they heal. It shifts the mental state from stressed to serene. It can inspire a conversation between the pediatrician and child. It is far easier, for example, for children to talk about their drawings than about their feelings (Malchiodi, 2009).

Children's art, according to experts, is filled with insights into kids' minds if you know what to look for (Buck, 1984). It is possible, for example, to identify what stage of graphic development characterizes a child's pattern of drawing (Roland, 2006). Four stages of graphic development are recognized between preschool and the upper elementary grades. Between ages one and two years, children scribble. The scribbles are at first random lines, followed by circles and geometric shapes. Studies have attempted to identify the different kinds of scribbles that are universal among children (Jarboe, 2002).

Children between the ages of three and four start to combine the circles and lines to create human-like figures. They are "tadpole" like that lack body parts, such as arms, legs, and body. Between the ages of seven and nine, the figures become more detailed. People figures are drawn with all the necessary body parts. At nine years of age, the child begins to develop even more detail in drawing people. Shapes now have recognizable forms with shadows and shading. People figures show varying expressions and attention is paid to clothing, pupils, lips, freckles, etc. (Roland, 2006). Because children's drawings can be segmented into specific stages based on the repeating patterns, it is possible to distinguish when a child is behind age level or in rare cases, such as autism, significantly ahead (Jarboe, 2002).

Therapists use a standard technique to discover how children perceive themselves in the world. It is called Draw

Figure 73. Therapists use a standard technique to discover how children perceive themselves in the world. It is called "draw a house-tree-person." The house mirrors the child's home life and family relationships; the tree reflects relationships that the child experiences within his or her environment; and, the person echoes how the child feels about herself or himself. Image courtesy of Shutterstock.

a House-Tree-Person (Buck, 1984). The house mirrors the child's home life and family relationships; the tree reflects relationships that the child experiences within his or her environment; and, the person echoes how the child feels about herself or himself. When children reach six years of age, this is the sweet spot when their drawings really matter. Visual patterns are clues to their relationship to family and friends and the environment and the image that children have of themselves (Dewar, 2014). When, for example, kids draw family members holding hands, it is a good sign of a friendly relationship among the family members. Drawings created with bright colors and happy faces suggest a sense of belonging. When parents are going through conflicts of some sort, the kids will draw the parents far apart. Kids may draw themselves distant from their parents or much smaller relative to their others in the family, when there is chaos in the home ranging from clutter to crowding. A child who has been abused might draw an abuser with bigger hands than normal (*How to Analyze a Child's Drawings*, 2013).

When children's drawings were first studied to identify patterns in child development in the 1880s, experts thought that cultural differences mattered little. Because the social world of friends and families is shaped by the culture of the child, we would, however, expect drawings to reflect that. They, in fact, do. Drawings from three different cultures were compared in a recent study; German upper-class families; Cameroon farming families, and Ankara, Turkey, urban middle-class families (Gernhardt, 2013).

Patterns differed in the depiction of selves and arrangement of their family members in several ways. Cameroon children drew families that included members other than their father or mother. The father figure was in fact mostly absent. This is consistent with Cameroon culture, which is characterized by an extended family of multiple caregivers and family. This contrasts with Ankara and German drawings that show only the father and mother. Children draw themselves next to their mother and father who are the principal caretakers. In Cameroon, it takes a village. Cameroon children often draw themselves next to nonrelatives. Ankara and German children often draw their figures smiling. Cameroon children do not. They are taught to control their emotions and stay emotionally neutral. It is not surprising that Cameroon children also drew the fewest facial features. It was equally interesting that Ankara children were the only group that drew eyebrows on their figures. It could be because eyebrows are important in the Turkey culture to communicate nonverbally. Tipping your head forward means "yes" but lifting your head backwards and raising your eyebrows means "no" (Gernhardt, 2013).

Figure 74. Some 7,000 children at the age of four were asked to draw a child. Figures were scored on the presence and correct quantity of human features. The more features pictured, the higher the score. Image courtesy of Wikimedia Commons.

The refrigerator door may serve as more than a gallery of children's art. It may predict a child's intelligence later in life. A child's intelligence at 14 years of age may be predicted by how well he or she draws at age four (Arden, 2014). Some 7,000 U.S. children at age four were asked to draw a child. Figures were scored on the presence and correct quantity of human features. The more features pictured, the higher the score. Seven years later, the same children were asked to take an intelligence test. The results from the intelligence test were then related to the results of the drawing test. A positive correlation was found, suggesting that drawings at age four could predict later intelligence (Arden, 2014).

Making the Invisible Visible

We can see only five percent of the universe. The remainder is dark matter. New visual tools, however, now permit us to recognize patterns that until now have never been seen.

Tools such as cell phones, Google Fusion, NASA snapshots taken from space, Wordle, and Microsoft Excel can uncover patterns once invisible and make them visible.

We, for example, have been taught that the orbit that the Earth takes annually around the Sun creates the seasons. Until we could look back and see the Earth from space, it was natural to think that our seasons resulted from Earth's changing distance from the Sun: the Earth is warmer in summer because we were closer to the Sun and colder in winter because we were farther away. The distance from the Sun during winter and summer, however, is just the reverse of what we might expect. That the northern and southern hemispheres have summer and winter at opposite times of the year offers a clue to the real reason for the seasons. The reason for the seasons is caused by the tilt of the Earth's axis away or toward the Sun as it travels through its year-long path around the Sun.

The Earth's tilt, however, is invisible from our perspective here on Earth. A NASA weather satellite, launched in 2010 to stay above the Earth at one site, captured the Earth's winter and summer positions from space. The amazing images below on the left show how sunlight fell on the Earth on December 21 and June 21—the winter and summer solstices, respectively, in the northern hemisphere.

During the summer solstice, the northern hemisphere tilts toward the Sun spreading more light and warmth to the north (*lower left* of Figure 75). On the winter solstice, sunlight spreads more to the south because the Earth tilts away from the Sun (*upper left* of Figure 75). Until the NASA satellite made the remarkable solstices images visible from space, we relied on models to explain the seasons. These images confirmed that the models were accurate. In addition to these patterns, many others are visible on Earth from space. Satellite images

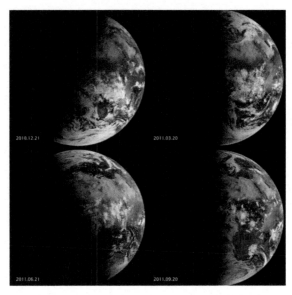

Figure 75. Image of Earth from top left during the 2010 winter solstice, 2011 spring equinox (upper right), summer solstice (lower left) and fall equinox. Image courtesy of National Aeronautics and Space Administration.

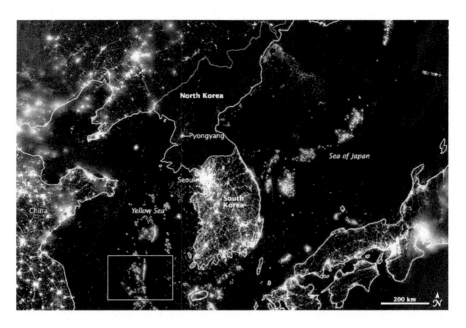

Figure 76. Korea at night. Note that in contrast to South Korea, North Korea is almost totally in the dark. Image courtesy of Wikimedia Commons and National Aeronautics and Space Administration.

that show the Earth at night reveal striking patterns. The image below differentiates unlighted North Korea from highly illuminated South Korea.

Sometimes, a pattern is present and nearby, but nearby noise masks the signal. This is especially true for number and word data for which there is sometimes so much data that it buries the pattern. A tool that mines large amounts of word data for patterns is the application Wordle. Wordle is in the business of drawing words. Designed first as a toy, one can paste a bunch of text into a window at the Wordle website. Click once and the application produces a beautiful word cloud that enlarges and color codes words that are used most often.

Jonathan Feinberg never imagined the uses for Wordle when he invented it (Roush, 2009). Wordles are published regularly in newspapers and on websites. Journalists use Wordle to research speeches, teachers to research literary works, and marketers to research customer surveys. I interviewed faculty and students as part of a yearlong project to research what great teaching looks like. The Figure 77 shows a Wordle constructed from a survey I did with first-year faculty at my college who had little previous teaching experience. New faculty repeatedly mentioned *lecture* as the best and only way to teach. The professors' main roles were to deliver knowledge, in their view. Engaging students was not a high priority.

Students when surveyed about great teaching favored professors who engaged them in learning and knew not only their subject, but knew the students as individuals. They respected professors who were well prepared, practiced good time management, and were enthusiastic about their subject. When interviewed, veteran faculty concurred with students

Figure 77. A Wordle constructed from a survey done with first-year faculty at Springfield College who had little previous teaching experience. New faculty repeatedly mentioned lecture as the best and only way to teach. Image courtesy of Robert Barkman.

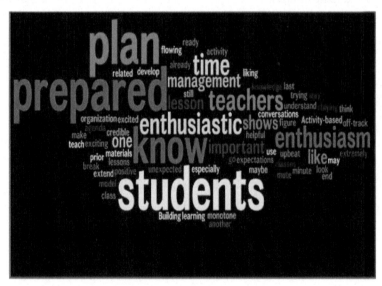

Figure 78. A Wordle constructed from a survey done with first through fourth year college students. Students, when surveyed about what great teaching looked like, favored professors who engaged them in learning and knew not only their subject, but knew the students as individuals. Image courtesy of Robert Barkman.

in many ways about what great teaching looks like. Engaging students was a high priority. They mentioned hands-on learning and using visuals to teach as important. A high priority was placed on showing students how theories could be applied to real-world situations. The data shows that new and inexperienced faculty considered what to teach more important than how to teach. Veteran professors recognized that engaging student in the subject matter was as valuable as teaching the content of the subject matter.

It is surprising to find that not all individuals are created equal when it comes to generosity. There are many ways to measure gifts of time and treasure, but measures of what percent of income that people give to charity is reliable and commonly used. When the percent of gross income that individuals give to charity is related to their gross income, a surprising pattern emerges. Poor and middle class Americans

Figure 79. A Wordle constructed from a survey done with veteran college faculty who had five or more years of teaching experience. Veteran professors recognized that engaging student in the subject matter was as valuable as teaching the content of the subject matter. Image courtesy of Robert Barkman.

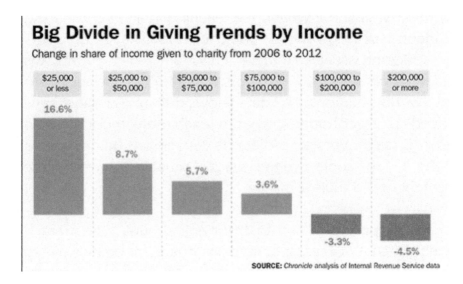

Figure 80. Poor and middle class Americans dig deeper into their pockets than wealthier Americans. Image courtesy of Chronicle of Philanthropy.

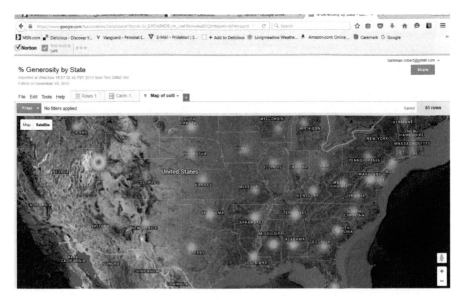

Figure 81. A state-by-state comparison done with Google Fusion that color codes each state according to the percent of income that state's citizens give to charity. Image courtesy of Chronicle of Philanthropy.

dig deeper into their pockets than wealthier Americans (*Almanac of American Philanthropy*, 2012).

The difference between the rich and poor gets more interesting when individual states are compared. A state-by-state comparison was done with Google Fusion that color-codes each state according to the percent of income that state's citizens give to charity. Generosity is measured by the size and the intensity of the color; the larger the size and the deeper the color, the more generous the state. The collective generosity of all the states is shown in Figure 81 as a "heat map." Like all heat maps, color is used to communicate data values that would be much more difficult to grasp if the data was presented numerically in a spreadsheet.

Utah is the most generous state with a giving rate of 6.6 percent. This is explained by the large number of Mormons who live in the state who are expected to tithe 10 percent of their income to the church. Mississippi, Alabama, Tennessee, Georgia, and South Carolina, where a high percentage of the states' populations are churchgoers, closely followed Utah. The most miserly states are the New England states. New England states suffer from low rates of church attendance exacerbated by Yankee independence and a tradition of self-reliance (Chronicle of Philanthropy, 2012).

Cell phones offer a new tool to study ourselves in ways that have never been studied before. Researchers are able to predict individuals' wealth from cell phone use. The graph below shows a strong correlation between actual wealth and wealth predicted from cell phone use (Blumenstock, 2015). This study was inspired by the paucity of demographic data available from third-world countries where censuses and household surveys are rare and unreliable. Reliable, quantitative data on the economic characteristics of a country's

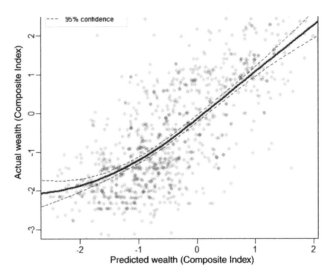

Figure 82. A strong correlation exists between actual wealth and wealth predicted from cell phone use. Image courtesy of Science.

Figure 83. The discovery of the cause of cholera is a story about a map. When plotted, the map showed where those who died (solid dark line) and those who survived cholera lived on Broad and the adjoining streets of London's Soho district. Note the proximity of the pump (solid circle) to the deceased. Image courtesy of Ralph R. Frerichs, UCLA Department of Epidemiology.

population, however, are essential for sound economic policy and research (Blumenstock, 2015).

In developing economies, where traditional sources of population data are scarce but mobile phones are increasingly common, these methods may provide a cost-effective option for measuring population characteristics. Whereas a typical national household survey costs more than $1 million and requires 12 to 18 months to complete, the phone survey conducted here cost only $12,000 and took four weeks to administer (Blumenstock, 2015).

Disease Patterns

The discovery of the cause of cholera is a story about a map. When plotted, the map showed where those who died and those who survived cholera lived on Broad and the adjoining streets of London's Soho district. John Snow discovered that someone could avoid contact with a person infected with cholera, but still contract cholera if one lived in the same neighborhood as its victims. Snow grasped that solving the mystery of cholera depended on reconciling these two seemingly contradictory facts (Johnson, 2006).

The missing link was water. The bug responsible for the disease is *Vibrio cholerae*, a comma-shaped bacterium that is transmitted by drinking contaminated water. It was no accident, as Snow was to learn, that many of the more than 700 victims who died over a period of two weeks lived clustered around the Broad Street water pump. The Broad Street pump was a reliable source of clean drinking water and was favored because of its mild carbonation. The seemingly high correlation between the cholera deaths and drinking water from the Broad Street pump was further strengthened by Snow's attention to outliers (i.e., people who did not fit the pattern).

Snow studied pockets of life where one would expect death and deaths where one would expect life (Tuthill, 2003; Johnson, 2006)

For example, in a brewery near Broad Street, none of the 80 workers fell ill from cholera. The drinking water in the brewery, Snow learned, was supplied privately, and workers favored drinking the brewery's fine malt liquor over water (Tuthill, 2003). Despite living at some distance from the Broad Street pump, several schoolchildren puzzlingly succumbed to the disease. It turns out that, on their way to and from school, several drank the pump's water. In an effort to close the Broad Street pump, Snow presented his data to the Broad Street board of governors. The board responded quickly and ordered the pump handle removed. The deaths from cholera quickly subsided. From Snow's careful observation of visual and social patterns, the cause of cholera was finally discovered. His approach to tracking the cause of disease became the foundation for the future study of epidemiology of disease (Johnson, 2006). It is what guided Allen Steere's own discovery of Lyme disease.

The history of Lyme disease reads like a detective story. The author takes readers on wild goose chases and down blind alleys before the cause of the disease is discovered. It is a story of visual patterns that connected when and where patients acquired the disease to how the disease was communicated. When Dr. Allen Steere and his colleagues finally fit the patterns together to complete the puzzle, the reader learns the importance of being observant, curious, and tenacious to solve medical mysteries like Lyme disease (Steere, 2016). The cause of the disease is a species of *Ixodes scapularix*, better known as the deer tick. It is also a story of how one women, who, through her passion to know, helped to solve the mystery of Lyme disease and add a new chapter to the annals of the history of medicine (Murray, 1996).

The story begins in the 1950s when Polly Murray noticed some strange symptoms about herself, such as swollen joints, rashes, and fever. She journeyed to doctor after doctor trying to find relief and the cause of her symptoms. She was diagnosed with a range of illnesses from lupus to chronic fatigue syndrome (Yankee, 2007). Later when her family moved to Old Lyme, Connecticut, her family, surprisingly, began to have the same symptoms as she did. The joint problems that her children experienced led doctors to diagnose them with juvenile rheumatoid arthritis. Like a good scientist, Murray kept good notes of the pattern and chronology of symptoms and treatments that she and her family shared (Murray, 1996).

A break in the case came when others in her community showed the same symptoms. At a dinner party, she learned that a friend's daughter came down with the same symptoms as she and her children suffered. She further learned that there was a high incidence of children diagnosed with juvenile rheumatoid arthritis in her school system. The high incidence of the disease in her family and community inspired Murray to seek help from the Yale University School of Medicine. Because it was unusual for an entire family to suffer from arthritis, it piqued the curiosity of a Yale University rheumatologist, Dr. Allen Steere (Steere, 2016). Guided by Murray's notes, Steere set out to find and evaluate all the residents of Lyme and surrounding communities who had symptoms of arthritis.

Steere's training in epidemiology prepared him to search for a pattern among the 39 children and 12 adults who were identified as having the symptoms of "Lyme arthritis." The normal incidence of the autoimmune disease is one of every

10,000 or 100,000 children in the population. He recognized that the incidence of arthritis of Lyme residents was 100 times higher than expected in the normal population (Steere, 2016). He asked, could this be a new disease? When those outliers with the disease were mapped, it showed that the disease was clustered. Half of those affected lived in wooded areas on two adjoining roads (Steere, 2016). The majority of patients noted that the onset of symptoms occurred in the early summer or early fall. The geographic and seasonal clustering of patients suggested to Steere that the disease might be transmitted by a tick. Coincidently, a few patients noted that until recently, few ticks were seen in the area. "Now they were inundated with ticks," and this was 1975 (Steere, 2016).

Some patients had a "bull's-eye"-like lesion around an area that looked like an insect bite. Some patients even re-membered being bitten by something before the onset of arthritic-like symptoms. This lesion fitted the description of an entity that is usually thought to be caused by a tick bite (Murray, 1996). Steere followed new cases of Lyme disease to try to connect the skin lesions associated with Lyme disease to the tick. It was probably no coincidence that 75 percent of the new cases showing a skin lesion developed arthritis. The smoking gun that Steere and researchers had sought was discovered when a field biologist doing a study in the Lyme area collected a tick that had bitten him. Several days later he developed symptoms of Lyme arthritis. The culprit was iden-tified as a species of *Ixodes*, the deer tick (Steere, 2016). The question, then, became, what is the disease-causing agent that the tick is transmitting? This was not so easy to answer.

Over a period of eight years, Steere accumulated evi-dence—more pattern recognition—to show that the agent was a spirochete bacteria. The bacteria was found both in the tick and the victims of Lyme disease. Like other spiro-

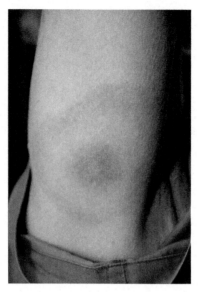

Figure 84. Some patients had a "bull's-eye"-like lesion around an area that looked like an insect bite, but it is really a bite from a tick (Arachnida). Image courtesy of Shutterstock and Anastasia Kopa.

Figure 85. The smoking gun that Dr. Steere and researchers had sought was discovered when a field biologist doing a study in the Lyme area collected a tick that had bitten him. Several days later he developed symptoms of Lyme arthritis. The culprit was identified as a species of Ixodes, the deer tick. Image courtesy of Shutterstock and Tomasz Klejdysz

chete-caused diseases, Lyme disease occurs in stages. Early symptoms include a rash, which may be followed weeks to months later by neurologic, cardiac, or joint abnormalities. Moreover, Lyme disease, like most other diseases caused by bacteria, responded to antibiotics. When their blood was assayed, everyone that shared this clinical picture of Lyme disease developed an antibody response to the spirochete bacteria (Steere, 2016).

Lyme disease is now the most commonly reported tick-borne illness in the U.S. The disease is spreading greatly and, in the Northeast, it is particularly virulent; people in one house after another living on the same street have come down with the disease (Steere, 2016). A doctor that Polly Murray consulted told her that some day she would be able to look back and see how the pieces of the puzzle of her mysterious disease fit together (Murray, 1996). Now, more than 40 years since Allen Steere launched his study of Lyme disease, she is able to do that.

The story of Lyme disease does not end there. First mistakenly diagnosed as juvenile rheumatoid arthritis, Lyme disease is one of the first autoimmune diseases known to be triggered by a bacterial agent. Autoimmunity occurs when the immune system recognizes and attacks host tissue. In addition to genetic factors, environmental triggers (i.e., in particular, viruses, bacteria, and other infectious pathogens) are now thought to play a major role in the development of autoimmune diseases (Ercolini, 2009). For the past 20 years, Steere and his colleagues have researched how the features of Lyme disease and other kinds of chronic inflammatory arthritis overlap, hoping to show that Lyme disease is not the outlier that people once thought (Steere, 2016).

The efforts of Lyme researchers to find a pattern among patients in space, time, and common circumstances closely followed the strategy John Snow used to discover the cause of cholera 150 years earlier (Edlow, 2003).

When I was a child living in Ohio during the late 1940s and 1950s, my parents lived in fear that I might contract polio. They prevented me from attending movie theaters and going to local beaches where crowds would gather. The fear was real. There was no cure for polio, which left its victims mildly to severely paralyzed. Ever since the outbreak of polio in June, 1916, the arrival of summer came with the fear of infection with polio. The worst epidemic was in 1952 in which 3,000 people died and 21,000 suffered some kind of paralysis (Aberlin, 2015). Thanks to the miracle of a vaccine developed by Jonas Salk, most have been protected from contracting polio since that epidemic (Wikipedia, n.d.). Except for in a few places scattered around the world, polio has been almost eradicated.

Disease Patterns: Course Corrections

Polio is just one of many infectious diseases that were common during the first half of the 20th century. More than a hundred years ago, communicable diseases such as typhoid, cholera, and influenza far outweighed our concern for non-communicable diseases, such as heart disease, cancer, and diabetes. What that change tells us is that causes of death have shifted dramatically in the past 100 years (Dietert, 2015). The pattern today has shifted away from deaths caused by infection to deaths caused by noncommunicable diseases. Noncommunicable diseases now account for 63 percent of mortalities. Risks in our environment now outweigh the risks of diseases communicated by agents of infection, such as bacteria and viruses. Obesity, smoking, and pollution are among our biggest concerns now... Hold on! That was until infections from COVID-19 were first reported in China in 2019 and changed the pattern of diseases.

The correction to the disease pattern in which the risk for disease is now viral underlies the current pandemic. The burning questions are when will the pandemic end and how will it end?

The option of vaccinating our way out of the pandemic is the ideal way. Except for a few countries, such as the United States and larger European nations, this is an expensive toll highway for most countries to access in the near term. Scientists can look, as well, to the past to help answer these questions and predict the future (Branswell, 2021). The pattern of past human coronavirus infections indicates that COVID-19 will, at some point, join a handful of human coronaviruses that become endemic, causing colds, mainly in the winter, when conditions favor their transmission.

Pandemics always end. Vaccines, except for ones that eradicated smallpox, have never played a significant role in ending them. That history doesn't mean vaccines aren't playing a critical role this time. Far fewer people will die from COVID-19 because of them. In 1918, when millions died worldwide from the influenza, there were no vaccines. When the H2N2 pandemic swept the world in 1957, the flu vaccine was mainly a tool of the military. By the time the United States produced enough vaccine, the worst of the pandemic of 1968 had passed (Branswell, 2021).

How did those past pandemics end? The viruses didn't go away. A descendent of the Spanish flu virus, which is the modern H1N1, circulates to this day, as does H3N2. Humans didn't develop herd immunity to them either. Instead, the viruses evolved over time, as did we. Our immune system adapted over time to learn how to fend them off. Moreover, the virus changed to a less virulent type. Instead of causing tsunamis of devastating illness and death, over time the viruses came to trigger small surges of milder illness. The viruses became endemic. Pandemic flu became seasonal flu (Branswell, 2021).

Cancer is the second leading cause of deaths in the United States. Smoking remains the leading preventable cause of death. The good news is that the number of smokers has declined since the Centers for Disease Control first started tracking them in 1965. The percentage of U.S. adults who smoke cigarettes in 2011 was 13.5 percent, a decrease from 42.4 percent in 1965 (Centers for Disease Control and Prevention, 2011). Credit for these decreases is attributed to tobacco taxes, ordinances that require smoke-free public places, and counter-marketing. These interventions are important tools, but how they accomplish their results is not clear.

One factor overlooked until recently is the influence we have on each other (Christakis and Fowler, 2008). That's why, for example, we are more likely to recycle if we see our neighbors recycle or patronize a store because our neighbor recommended it (Badger, 2014). Can the same factor influence smoking cessation? Recent results show that it can. A study carried out by Christakis and Fowler showed that social ties between spouses, siblings, coworkers, and friends can influence each other's smoking behaviors (Christakis and Fowler, 2008).

The subjects for their study were more than 12,000 subjects who took part in the Framingham heart study initiated in the 1970s. They studied smoking behavior among the social network for more than three decades (Aubrey, 2008). The overall smoking behavior decreased over 3 decades paralleling the decrease, which was observed nationally. The findings suggested that the decision of many to quit smoking is not made in isolation but represents choices made by groups of people connected directly or indirectly. Experts observed further that the social network became more polarized over time; smokers clustered with smokers and nonsmokers clustered with nonsmokers. Smokers occupied the center of their circle of friends early in the study, but by 2000, smokers were marginalized at the periphery where they were connected to only one or two people (Christakis and Fowler, 2008).

The probability that a subject would quit smoking was contingent on the type of relationship the subject had with the nonsmoker (Aubrey, 2008). Among married couples, when one spouse quit, the chances that the other spouse would quit was 68 percent. Among close friends, the probability was 43 percent. Someone's education background also made a difference. The higher the education achieved, the more likely friends were to successfully pressure each other to quit (Christakis and Fowler, 2008). The take-home lesson from the patterns observed seems to be that a more productive approach is to target an individual's social network rather than the individual. Alcoholics Anonymous and other institutions that practice group therapy, of course, realized this long ago.

Surprising Ways We See the World

Alexandra Horowitz's *On Looking: Eleven Walks With Expert Eyes* shows us how to see the ordinary in new ways (Horowitz, 2013). *On Looking* is structured around a series of 11 walks the author takes, mostly in her Manhattan neighborhood, with experts on a diverse range of subjects, including an urban sociologist, an artist, a geologist, a physician, and a sound designer. She also walks with a child and a dog to see the world as they perceive it. What they see, how they see it, and why most of us do not see the same things reveal the startling power of human attention to patterns and what it means to be an expert observer. It was an eye-opener for Horowitz (Horowitz, 2013).

It was an eye-opener for me years ago when I observed a student teacher teaching a lesson on world geography to high school freshmen. To discover how much students already knew about world geography, the student teacher asked them to draw a map of the world from memory including mapping their home state. Their mental maps of the world got our attention. Most of the student maps surprisingly showed North America centered in the middle of the map and drawn markedly larger than the other continents. In many maps, whole continents and countries were missing. Their home state of Massachusetts occupied almost the entire New England region. This paralleled a study that asked 30 college undergraduates to draw a map of the world from memory. When the maps were merged together into one image, it showed North America centered on the paper. Antarctica, the Arctic, Greenland, New Zealand, Madagascar, Scandinavia, the British Isles, and "most of Southeast Asia" were missing (Friedman, 2014).

In a study called "High Schooler's Image of the World Before Mapping the World by Heart," students are asked to draw a map of the world, with no help from textbooks or reference materials. Most of the maps are sketchy blobs and squiggles. After Mapping the World by Heart, the same students could draw, entirely from memory, world maps that include the names, borders, mountain ranges, rivers, and cities of over 150 countries—all properly proportioned and correctly situated by latitude and longitude.

Figure 86. New Yorker's view of the world. This tongue-in-cheek view of the world illustrates a New Yorker's mind view of the rest of the world; the entire world is a suburb of Manhattan. Image courtesy of The New Yorker.

Figure 87. Close your eyes. Picture a scientist in your mind's eye. What is the scientist wearing and holding? Were you imagining Albert Einstein? Image courtesy of Shutterstock.

We all probably harbor idiosyncratic views of the world that overemphasize parts we favor and ignore others we know less about. This view is reinforced by the famous cartoon cover of the March 1972 issue of *The New Yorker* magazine voted among the best magazine covers ever created (Wikipedia, 2012). New York between 9th Avenue and the Hudson River is depicted taking up half the world map. The rest of the country between New Jersey and California is barren desert. The Pacific Ocean separates the United States from small renderings of China, Japan, and Russia. This tongue-in-cheek view of the world illustrates a New Yorker's mind view of the rest of the world; the entire world is a suburb of Manhattan. Every time I see an "I Love New York" bumper sticker, it reminds me of New Yorker's view of the world.

Close your eyes. Picture a scientist in your mind's eye. What is the scientist wearing and holding? Describe the facial features and hair. What kinds of equipment are in the picture? Now, open your eyes and, quick, draw the scientist! Does the scientist have wild hair and wear glasses? Is the scientist wearing a lab coat and working in a laboratory surrounded by test tubes, beakers, and a microscope? Were you imagining Albert Einstein? After you have finished, ask yourself, is the scientist male and White?

If you answered yes to most of these questions, your mental image agrees with the visual pattern most people picture when drawing a scientist (Symington, 2006). What is wrong with this picture? The answer is that it stereotypes scientists who are not necessarily male, White, or work in a laboratory. It confirms what two anthropologists confirmed more than 50 years ago when they tested how more than 3,500 high school students pictured a scientist. The most popular image was a study in white and maleness. By the fifth grade, the stereotype of a scientist is already present and persists through adulthood (Finson, 2014).

We then wonder why so many girls and nonwhite children find it very difficult to envision themselves as future scientists (*GeekFeminism.org*, 2010). The data about women and science supports this. In 1970, women made up 7 percent of the science, technology, engineering, and mathematics (STEM) workforce. By 1990, there was an uptick to 23 percent. The rise stopped there. Today, it is about 26 percent (Del Giudice, 2014). Researchers do not know, what, if any, influence stereotypical images have on shaping children's perception of science, but the indications are that negative stereotypical images translate into negative perceptions of science (Finson, 1995).

The view of scientists as older White men in lab coats presents a highly exclusive view of who can be scientists. In recent years, the push to increase girls' and women's interest in science has come from seemingly everywhere, from the federal government to toy companies (think GoldieBlox and Computer Engineer Barbie) (Weinstein, 2014). The good news is that the effort to change the perception of "man as scientist" is working in small, but significant ways. One effort is called Girls in Science Camp. A really exciting moment in the Girls in Science Camp is the revelation of how drastically the girls' perceptions of scientists change from the beginning to the end of the week.

At the beginning of the week, the girls were asked to draw a picture of a scientist, and many of them drew a "mad scientist" with hair standing on end, long lab coats, and brightly colored potions. By the end of the week, they were asked to draw a scientist again, and this time all of the girls drew women in traditional lab settings, at home, with pets and children, or doing their favorite hobbies. Their pictures showed how their perception of a scientist expanded from the male, mad scientist and diversified into intelligent, multifaceted women with many interests and real lives. Their realization came because of interacting with a diverse range of women scientists and learning about their lives in and outside of lab (Roach, 2016).

These positive outcomes have been reinforced in school settings through teacher intervention programs designed to change teacher practices. Teachers have been encouraged to create a gender-neutral classroom, invite women scientists to serve as role models, and engage students in hands-on learning (Mason, 1991). Let's hope we will not be surprised in the future by images of women when students are asked to draw a scientist.

Seeing is much more than having our eyes wide open. We are surprised sometimes by how little we see of the world

Figure 88. Sister scientist. Their pictures showed how girls' perception of a scientist expanded from the White male, mad scientist and diversified into intelligent, multifaceted women of color with many interests and real lives. Image courtesy of Shutterstock.

around us when our eyes are wide open. Stick out your thumb. The thumb is the amount of detail that you can see well, a small subset of our visionary field. The area around the thumb is fuzzy from lack of detail. However, it is not the eyes, but the brain that limits what we see. When our attention is focused on one thing, we fail to notice other, unexpected things around us—including those we might want to see (Simons, 2012). Most of us have experienced pulling out of an intersection and, because of a blind spot, overlooking an oncoming car. You can understand this firsthand by watching an experiment (below) conducted by Daniel Simons and Christopher Chabris called the Monkey Business Illusion (Simons, 2010, April 28).

The experiment is designed with two teams of three members each. The members of one team are wearing white and the other team members are wearing black. Each team passes a basketball back and forth as they move and weave around each other in a circle. Your challenge is to count the number

Figure 89. Viewers of this video were asked to count how many times white-shirted players passed the ball. Fifty percent of them didn't see the woman in the gorilla suit. Image courtesy of Simons, D. J., & Chabris, C. F. (1999). Gorillas in our midst: sustained inattentional blindness for dynamic events. Perception, 28, 1059-1074. To learn more about this fascinating gorilla study, visit www.dansimons.com or www.theinvisiblegorilla.com. Image courtesy of Daniel Simons.

of times the white team passes the ball back and forth. View the video (Simons, 2010). The video can be viewed at either of these websites: *www.theinvisiblegorilla.com* and *www.dansimons.com*.

How many times did the team in white pass the ball? Did you count 16 times? The number is not really important. The real purpose of the experiment was to learn how much we see outside our field of focus. Halfway through the video, a person dressed up like a gorilla walks on stage, stands in the middle of the group, beats its chest, and walks off. It is remarkable that half the people that view the experiment overlook the gorilla (Simons, 2012).

We are not the masters of attention that we think we are. We are often surprised by events happening right in front of our eyes that we overlook. It could be the gorilla in the middle of the room, a symptom of a disease that was there all the time, or a business expense that we ignored.

Millions of Americans have no savings set aside for a rainy day, leaving them in serious jeopardy if financial calamity strikes, according to two recent studies. Roughly one-third of American adults do not have any emergency savings, meaning that more than 72 million people have no cushion to fall back on if they lose a job or must deal with another crisis (Jones, 2015). In the last 30 years, the savings rate of 90 percent of U.S. citizens has fallen from six percent to negative four percent (Thompson, 2014). The United States is the outlier compared to our European and Asian counterparts. The latter countries' citizens are much better savers and more likely to save more money for retirement than Americans (see pattern below relating savings rate to country). The Germans and the French save over 12 percent of their income, the Japanese

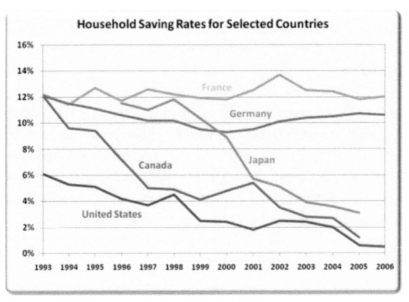

Source: Organization for Economic Cooperation and Development (2007)
OECD Economic Outlook: December No. 82 – Volume 2007 Issue 2

Figure 90. Household savings rates reported for five world countries. Compared to the United States, France, Germany, Japan, and Canada are far better savers than the United States Image courtesy of Organization for Economic Cooperation and Development (OECD).

nearly 30 percent, and the Norwegians save nearly a whopping 35 percent of their income (Thompson, 2013).

The COVID-19 pandemic created a tale of two economies: Those who were able to save, and those who struggled to make ends meet. As the U.S. economy begins to recover and reopen, many consumers are still scrambling to regain their financial footing. While the conventional wisdom is to sock away 6 to 12 months' worth of savings, that became impossible for many during the COVID-19 pandemic, as millions of people lost their jobs, small businesses were forced to shutter, and day-to-day living expenses piled up. The stimulus checks helped, but not necessarily enough.

There was, however, some good news. As more Americans got vaccinated and infection rates eased, the U.S. economy slowly reemerged. Businesses reopened, hiring rose, and that eventually eased some of the financial strain felt by many.

In March 2021, the personal savings rate—which reflected the ratio of total personal saving minus disposable income—surged to 27.6%. While saving was up, that figure also indicated a short-term slowdown in consumer spending, as people held onto more of their money. The last time the savings rate was this high was April 2020, when it hit 33%. While it has slowly eased during the past 12 months, it has remained above 12%, compared with pre-pandemic levels that were below 10% (Tymkiw, 2021).

Savings rates and the amount we plan to save for the future are affected by a multitude of factors, such as income level, education, and religious affiliations. But what about language? According to Keith Chen, the structure of language shapes our judgment and decisions about the future. Languages, for example, differ in the ways they encode time (Chen, 2013). World languages are generally divided into futured and futureless languages. English, for example is a futured one. If we say that,

"I will save some more for retirement,"

Germans would say, "Ich etwas mehr für den Ruhestand zu sparen." This translates into,

"I save some more for retirement."

When English speakers talk about the future, the language requires that future time be marked by "will." This has the effect of making the future feel more distant. By disassociating the future from the present, people are more likely to procrastinate about saving for the future. The German language is futureless; it does not require future markers. Hence the present and future feel very much alike and close in time. Chen proposes that those speakers would be more willing to save for a future which appears closer (Chen, 2013).

Chen mapped stronger and weak future-tense languages across Europe and correlated the data with future-oriented behaviors like savings (Thompson, 2013). Having a larger proportion of people speaking languages that do not have obligatory future markers makes national savings rates higher. After factors such as education, income level, and religious affiliations were controlled for, the effect of language on people's savings rates turned out to be significant. By speaking a language that has obligatory future markers, such as English, those people are 30 percent less likely to save money for the future (Thompson, 2013). The answer to how future retirees can save enough for retirement seems simple enough—learn to speak German!

CHAPTER 6

PUT YOUR PATTERN FINDER TO WORK

It Takes an Amateur

As students of the history of science know, at one time, almost all scientists were amateurs. Copernicus was a monk, Mendel performed his experiments in a monastery, and Einstein was a patent clerk in Switzerland (Smith, 2012). By taking lessons learned from science, amateurs too can make discoveries small or large that benefit one's self, family, workplace, or an entire population of people. Each one of us has in him or her the power to make a new discovery by standing on the shoulders of others. It doesn't take a formally trained scientist to make groundbreaking scientific discoveries. Amateur scientists have contributed many. Amateurs can play key role in collecting and analyzing data. Motivated to take part in real research, anyone can participate, from youngsters who have an interest in bird-watching to retirees who are good at numbers.

An early example, launched in 1999, is the Search for Extra-Terrestrial Intelligence (SETI) Project (SETI, 2016). The project's mission is to listen for signs of intelligent life originating from outer space. The "ear" that listens 24/7 for signals is done with the world's largest radio telescope located at Arecibo, Puerto Rico. Because the amount of data gathered from listening is too great for any one computer to analyze, the data is analyzed by millions of individual computers strung together on the Internet (The Planetary Society, 2016). My computer was one of those. Anyone with a home computer can participate. When a computer is not in use, a screen saver jumps into action to get a chunk of data from over the Internet, analyzes that data, and then reports the results. When you need your computer back, the screen saver instantly gets out of the way. Amateur scientists, however, are usually engaged more actively in doing science. It can be a team approach where experts and nonexperts collaborate to plan, collect data, and analyze data for patterns. They can even publish and present papers together. Some even make game-changing discoveries together.

When I have a few minutes, I drop in to help my Galaxy Zoo co-researchers classify the shape of unknown galaxies. I am one of 250,000 amateurs, affectingly called "zooites," from countries around the world who are engaged in the same task. When I log on to Galaxy Zoo, I am shown a photo of a galaxy online, perhaps millions or even a billion light years away taken by telescopes around the world, including the famous Hubble space telescope (Galaxy Zoo, 2016). I am prompted to answer questions, such as, "Is the galaxy smooth and rounded or is it disk shaped?" I click on the shape that best fits the galaxy. The question is followed by others asking if a spiral is present, whether the spiral is loose or tight fitting, and how many spiral arms there are (PC Plus, 2011). As my skill to classify galaxy patterns improves, the time to classify them shortens: That is, until something unusual pops up on the screen. It could be two galaxies merging or colored objects we cannot identify that seem to form part of the galaxy. When an image grabs my interest, I take time out to comment on it. I anticipate that another zooite looking at the same image will either confirm that it warrants a further look or conclude that it is an artifact. Who knows? I may be on to something new!

Groundbreaking new discoveries do happen. Just a week after schoolteacher Hanny van Arkel had been looking at galaxies, she noted a startling green cluster on one image. A few other zooites had looked at the same picture, presumably not knowing what to make of the cluster and moved on. van Arkel put a note on the message board, and wondered if anyone knew what the bright green cluster might be. It became "Object of the Day," an ongoing series. But in this case no one, not even Kevin Schawinski, had ever seen anything like it before. "Hanny's Voorwerp" (object) quickly became a cause célèbre in the wider astronomical universe (PC Plus, 2011). And after much debate it was suggested that the *voorwerp* was indeed unique—it captured the moment that a quasar, a light beacon powered by a black hole, illuminated a gas cloud. The "quasar mirror," the size of the Milky Way, could yet provide an explanation of one of the more mysterious processes in the universe. Despite being only a few years old, Galaxy Zoo has led to many published peer-reviewed papers and claims of new galaxy discoveries.

Partner With a Professional

Whether you have your sights fixed on coauthoring a paper for publication or are motivated by the excitement of making a new discovery, you can choose from many projects that recruit partners. The project that best matches your interests and talents to the needs of the project is likely one that draws on your strength in recognizing patterns either in people, words, numbers, or nature or projects requiring visual-processing and spatial-reasoning skills. Here are some project choices to consider.

Discover New Patterns in Nature

If searching for new planets is your thing, you can detect new exoplanets in your own backyard with Planet Hunters TESS (*www.zooniverse.org/projects/nora-dot-eisner/planet-hunters-tess*). It taps into the power of human pattern recognition (NASA Ames Research Center, 2013). You, working as a planet hunter, use the interface on the Planet Hunters website to find the points where star brightness changes. Transit events—a brief changes in a star's brightness—occur when planets pass in front of a star. Finding a dip in a star's spark could point to a new planet (Braccini, 2012)! NASA scientists have confirmed 42 alien planets found by amateur astronomers. Fifteen out of these 42 alien planets may be habitable because they lie within the so-called Goldilocks Zone around a star, a limited range of orbiting distance considered to be optimal for life (Frye, 2013).

"ZomBee Watch" is a project that asks volunteers to track the behavior of honeybees that have been infected with a parasite. The typical zombie scenario—humans catch virus; humans infect other humans; and survivors fend off infected—has its counterpart in honeybees. In this case, a parasite changes bees' behavior, causing them to leave the hive at night and die. This has potential impact on agriculture, so investigators and amateur volunteers are mapping and analyzing the distribution of this zombie-bee behavior through *www. zombeewatch.org/theproject#.YRXndN8pCUl* (A Citizen Science Project Tracking the Honey Bee Parasite Apocephalus Borealis, n.d.).

Even your backyard can serve as a laboratory. Millions of acres of bird habitat the size of West Virginia and Maryland combined are lost each year to residential landscape and backyards. To adapt to this loss, investigators want to discover what makes a good habitat for birds and share this information with likeminded homeowners. "Yard Map" coaches you to make landscape maps of your yard and other green spaces like parks and nature reserves. Through *https://feederwatch.org*, amateur scientists can share their maps and collaborate with others to learn about what best works to improve habitat and attract more birds (The Cornell Laboratory of Ornithology, 2011).

Interested in astronomy? You are invited to join thousands of others observing the sky for new galaxies and planets. Galaxy Zoo (Galaxy Zoo, n.d.) invites citizen scientists to help classify photos taken of far-off galaxies by the Hubble Telescope. Conceived by an astronomy student overwhelmed by the number of photos he needed to classify, he reached out for help from amateurs with the Galaxy Zoo website he created (*www.zooniverse.org/projects/zookeeper/galaxy-zoo/ about/research*). In the first hour of launching the site, it recorded 70,000 classifications per hour. With a quarter million "zooites" now taking part, it is perhaps, the largest crowdsourcing project in existence. "It has created the world's most powerful pattern-recognizing super-computer that exists in the linked intelligence of all the people who had logged on to the website" (Adams, 2012). Despite being only a few years old, many peer-reviewed papers have been published and discoveries of new galaxies have been claimed through Galaxy Zoo.

Discover New Visual Patterns

If you're good at recognizing visual patterns and spatial reasoning, consider joining the project Foldit (Xue, 2014). The "it" of Foldit is a protein. Proteins are the sub-microscopic molecules that are composed of hundreds, even thousands, of the 20 naturally occurring amino acids linked together in chains. The arrangement of amino acids (that is, the order by which the 20 different amino acids are linked together) is coded for a cell's DNA. When a newly synthesized protein peels off DNA as a linear chain of amino acids, it does not remain linear very long. It self-folds simultaneously into a three-dimensional (3D) structure whose shape characterizes each of the 100,000 proteins that make up a human cell. The 3D shape decides what the protein does. Some proteins like hemoglobin are designed to carry oxygen; others are shaped to form antibodies that lock on to viruses or enzymes to increase the rate of biological reactions.

What's amazing is that out of the millions of ways that a protein can be folded, only one pattern works. That's where Foldit comes in (Xue, 2014). Two computer scientists, Seth Cooper and Adrien Treuille, set out to discover the one-in-a-million ways to fold a protein using the power of crowdsourced computers. The computer program that they created, called *Rosetta*, was adapted to run as a screen saver that online volunteers could install in their home computer. Volunteers could watch the screen-saver experiment with the different ways to fold the protein, but could not interact with it. Some volunteers claimed that their intuition told them that they could do a better job than the computer. This inspired Cooper and Treuille to combine that computer power with human problem-solving abilities (Bohannon, 2009). That was the motivation to create the game Foldit.

Foldit players start with a partially folded protein structure and must manipulate its structure by clicking, pulling, and dragging amino acids until they've arrived at the protein's most stable shape (Xue, 2014). The more stable the structure, the higher that players can score (Keim, 2010). Researchers have shown that the humans' pattern-recognition and puzzle-solving abilities make them more efficient than existing computer programs at pattern-folding tasks. Most of thousands of people who have played the game have little to no biochemistry background and play the game in their spare time. They are a mix of at-home individuals, middle-school students, and people who just want to contribute. Some have been authors on four scientific papers, and their game play has contributed to several more publications (Xue, 2014). Together they solved the folding pattern of a viral protein that causes AIDS in weeks that computers took a decade to solve. Try it yourself: _https://fold.it/portal/info/about._

Discover New Number Patterns

Do you remember those contests in which you were challenged to guess the number of M&Ms in a jar? Or "guesstimate" the number of people who were at the ball game last night? Your gut number sense is your intuitive ability to quickly estimate quantities and their relationships. This skill evolved over eons when our ancestors were faced with deciding whether to take flight or to fight when attacked by an enemy. A mistake in guesstimating the size of the enemy could lead to an unfortunate ending. Research reveals that number sense is basic to all animals, not just human beings. For instance, creatures that hunt or gather food use their number sense to ascertain where they can find and procure the most nuts, plants, or game, and to keep track of the food they hunt or gather. Even newborn infants have it.

You can test your own skills at estimation and contribute to a crowdsourcing experiment sponsored by John Hopkins University researchers. Panamath measures your Approximate Number System (ANS) aptitude. It works like this. You are shown in brief flashes a series of slides with yellow and blue dots. Panamath asks you decide whether there are more blue or yellow dots. The simple task of deciding whether there are more blue dots or yellow dots in a brief flash can inform researchers a lot about the accuracy of your basic gut sense for numbers. Some comparisons are easy like comparing 6 blue dots to 12 yellow dots. Others are much harder like comparing 12 yellow dots to 14 blue dots. Because this task is so simple, Panamath can be used across the entire lifespan from two-year-old children to older adults. Here is a link to Panamath *http://panamath.org.*

A surprising pattern has been found between the ability to estimate numbers and mathematics skills. A recent Panamath study done of four-year-old children points to a strong positive correlation between the two skills. Recent research has demonstrated a relationship between performance on this test and basic mathematical ability. A child who has a good ability to approximate numbers, as tested by Panamath, also scored high on a standardized test of early math ability (e.g., numbering, number comparison, number literacy, calculation skills, and numbers concept). It is intriguing to speculate why the link between the ability to guesstimate numbers and math skills is present before a child begins formal education. It has not escaped researchers' attention that by raising children's abilities to estimate numbers, it may be possible to raise children's math skills. Think how you could put the resource of Panamath to work to partner with a school or youth organization or to raise the math scores of your own children.

Patterns buried in a mountain of numbers data are waiting to be discovered. There is much data available along with the tools to analyze it online. Join others to become part of a growing group of amateur data scientists. Your first stop should be at Kaggle (*www.kaggle.com*). Kaggle sponsors contests that partner amateur data scientists with major corporations to solve real-world problems (Jonas, 2014).

State Farm Insurance Corporation sponsored a competition to test whether data gathered from dashboard cameras can automatically detect drivers engaging in distracted behaviors. The winner was to receive a cash award of $30,000. In another competition, the online reservation company Expedia is challenging Kagglers to analyze its customer data to predict where a user will stay at 100 different hotel groups it advertises. Using 20 years of data of the late basketball great Kobe Bryant's swishes and misses, contestants are challenged in a just-for-fun contest to predict which of Kobe's shots will find the bottom of the net. Find contests like these at Kaggle, along with guides of how to get started in data science and tutorials for people who may not have a lot of experience in data science and machine learning.

Discover New People Patterns

Can you imagine what fighting in World War I must have been like? It must have been hell to sleep in a trench, with gnawing rats, itchy lice, and the scream of German shells launched above you. Private Harry Drinkwater, a former shop assistant, reckoned he was getting "approximately an hour's sleep a day." A new project is making it possible for people to see war through the eyes of World War I soldiers who recorded their experiences in diaries. The soldiers' diary entries exemplified by the following entries were put online as part of the centennial anniversary of the start of the war (Shapiro, 2013). Workers at the British National Archives have been busy scanning and uploading the documents to a website (Leggett, 2014). These accounts of soldiers' lives are especially important now because there are no longer any living British veterans of World War I (In the News, 2014). Learn more at *http://blog.nationalarchives.gov.uk/blog/operation-war-diary-archive-needs*

There was nothing "great" about the "Great War"—later named World War I. People of all ages studying this period of American history will agree after reading about the grim reality of soldiers' lives, as well as the lives of civilians, during wartime (Kids Discover, 2016). The secret diaries of a World War I British Army officer tell of life in the trenches, rats, death, and a longing to return home. There are people patterns waiting to be discovered (Leggett, 2014).

NOVEMBER 29, 1915

It (trench) has teemed with rats. The trenches are ankle deep—some places calf deep—in mud and the communications trenches are rushing streams of brown water.

The men are wet through but stick the job like Britons and I hope for their sake that the weather lifts with the morning.

DECEMBER 1, 1915

It is exciting work, sniping (the enemy). In fact, one must curb the tendency lest it should become a fascination.

One fellow was walking across the open—2,000 yards off—when I spotted him and let go. You never saw a chap move quicker in your life.

JANUARY 13, 1916

Three times I had to fling myself down in the wet grass, bury my nose in it and grovel while the [machine guns] went chattering over me. It is remarkable with what speed one learns to "adopt the prone position".

I long and long to see you, to clasp you in my arms... and I long with all my heart to see my Baby. How I love her. What hopes I have for her, what a sweet girl she will make.

Approximately 1.5 million pages of war diaries cover activity on the Western Front. The pages are full of fascinating details about the people involved and descriptions of their activities. The British National Archives is launching a crowdsourcing project called Operation War Diary. It asks "citizen historians" to help tag data, such as World War I people, places, dates, and activities (In the News, 2014). From the tags, they can create a detailed index to the people who appear in these pages and learn more about what they were doing (Leggett, 2014). A short tutorial walks citizen scientists through the classification and tagging process.

There are many people patterns to explore (UNICEF, n.d.). Civilian fatalities in wartime, for example, climbed from 5 percent at the turn of the 20th century to 15 percent during World War I, to 65 percent by the end of World War II, to more than

90 percent in the wars of the 1990s. How we can account for this the trend in civilian casualties? What were some of the new developments used for the first time in World War I? How did the new use of machine guns, poison gas, and airplanes change how people fought wars? Which of these developments continue to be used in battles today? Imagine soldiers from Britain, America, France, and Germany, each telling the same 100-year-old story from their own point of view (Shapiro, 2013). Would they follow similar or different patterns?

Picture yourself sitting in a waiting room passing time looking at an aquarium alive with fish and other creatures. What grabs your attention? Do you focus on individual fish or other creatures? Or, do you step back and note the bigger picture, perhaps, noticing how fish interact with each other and other creatures? It probably comes as no surprise that perceptions will vary among individuals. What is surprising is that perceptions may vary according to differences between East and West cultures.

When shown a short video clip of an underwater scene with fish, plants, and rocks, Americans reported they homed in on the brightly colored, rapidly moving fish. Japanese viewers differed. They paid more attention to the interactions that occurred, such as between fish and plants or fish and other fish. In contrast to Americans whose perceptions were absolute and independent of the environment, Japanese perceived things more in context; they stepped back to note the bigger picture (Kitayama, 2008).

Researchers followed up these observations to confirm this pattern of cultural differences with a simple online test. You are invited to take the same test to test the hypothesis of East-West differences at this web address. *www.labinthewild. org/studies/frame-line*

You will be shown a square frame with a vertical line. After being given a short time to memorize the line's length, you will be shown a new empty square. You will be challenged to draw a new line identical in length to the original line. Accuracy is judged between two ways of drawing the line. Draw the line both in identical length and length relative to the height of the box (Reinecke, 2016).

Which were you better at estimating—the absolute or relative length of the line? Although there is some controversy among researchers about the results, the concept of absolute versus context has applications beyond cultural differences. Knowing the difference between the two can help you make more informed decisions about the kind of risks to take. Some risk is something that we all expect to take in our lives. The amount of risk we choose to take varies depending on how the risk is expressed. Let's say the absolute risk of developing a certain disease is 4 in 1,000. If a treatment (e.g., exercise) reduces the relative risk by 50 percent, it means the 4 is reduced by 50 percent. Therefore, the treatment reduces the absolute risk from 4 in 1,000 to 2 in 1,000. This is not much in absolute terms (Newson, n.d.).

Along with these examples, there are many other opportunities to partner with professionals who study people patterns. The online resource, Labinthewild (*www.labinthewild.org*), offers a variety of ways to cooperate in studies of human behavior. It's a win-win situation. In exchange for participating in the study, participants are provided with personalized feedback that compares their results to others. Researchers benefit from including a greater number of subjects and more diverse population of subjects in their study. In their first two years of existence, the site was visited two million times and approximately 750,000 people participated in a study.

Some of the research includes studies that test your emotional IQ. Participants are shown just the eye regions of a face and asked to judge the person's mental state; for example, are they sad, angry, or happy? This research will help answer important questions about social behavior. Are women better judges of someone's emotional state than men? How do people of different ages compare? How well do people with autism judge someone's mental state? Another experiment tries to predict ages of participants based on how well they distinguish different colors. "Do you have the reaction time of a cheetah?" tags a study that measures how quickly someone notices visual changes. Yet another study tests your thinking style and how it compares to others.

Discover New Word Patterns

In which part of the world would you find yourself if you heard someone say, "It's angry you will be" or "It's myself that wants it?" It's unlikely that you guessed Newfoundland. The difference between the English that North Americans speak and that Newfoundlanders speak is their dialect. Newfoundlanders, in fact, would find the aforementioned sentences grammatically correct. There are 375 million people from around the world who speak English and that is just one of the hundreds of different dialects of English.

Researchers from the Massachusetts Institute of Technology are interested in how these different English dialects are affected by demographic variables (Hartshorne, 2014). They are trying to answer a two-part question. Is there a pattern to the dialect that someone speaks, where they live, and the age that they first learned English? You can partner with researchers and help answer these questions by taking an online quiz. Find it at _www.gameswithwords.org_. Your results along with others will help researchers predict the dialect of English you speak and your native language. I took the test and the results accurately predicted that my native language is English and my dialect is American (Standard).

Language is complex. Its complexity is what makes us human. Language is a product of thousands of years of evolution from when we first learned to communicate by mimicking the sounds of primates and birds to now. It has challenged linguists to understand words and their meaning. For example, how do we know when to use one word in a sentence and not another? Is the choice random or is there a pattern? It would seem simple enough to look up words in a dictionary to find their meaning. But, when you do, you find a definition, which itself is explained by more words that requires more searching for their definitions and so on (Hough, 2014).

The meaning of language is in the mind of the beholder and interpreted by the context of the sentence. Steven Pinker states that language emerges from human minds interacting with one another (Pinker, 2005). It is a device for moving thought from one person's head into another's. To understand language is to understand human thought. For example, if I use 'hooking up" in a sentence, it could mean that I am linking the parts of a computer together to be able to use it. Or, alternatively, it could mean that I am hooking up with another person to form a new relationship with that person. Someone cannot decide which meaning is correct simply by looking it up in a dictionary.

Through the VerbCorner project (*https://blog.scistarter. org/tag/linguistics/#sthash.Uo34zy61.ltnBLmMP.dpbs*), professional scientists and amateurs are working together to determine how grammar and meaning interact, and through that effort opening a window into the workings of the human mind. To answer this question, scientists are targeting the meaning of verbs. Why verbs? Verbs, according to these scientists, are the heart of both meaning and grammar (Hough, 2014). Steven Pinker calls verbs the chassis of a sentence (Pinker, 2005); it is the framework to which other parts of the sentence (e.g., subject, object) are attached.

Some linguists believe that, if you know the meaning of a verb, you can predict what its grammar will be like. There are practical reasons to know the relationship between grammar and word meaning, as well. For example, knowing about grammar and word meaning will make it possible to explain grammar to a child or to program a computer to recognize human speech in a meaningful way. The project is divided up into a series of seven tasks, each of which focuses on a particular kind of verb use and meaning. Because the exact meaning of a verb depends on the sentence it is in, each verb needs to be tested in all the sentences that it can appear. The results so far suggest that there is a pattern to verb usage. To continue this project, researchers need your help. As of 2016, 10,000 amateur volunteers have done more than a half million analyses (Hough, 2014). Help rewrite the rules of grammar by partnering with the researchers at VerbCorner. VerbCorner invites citizen scientists to answer fun questions about words and their meanings to eventually help train computers to understand language.

What Pattern Finders Have in Common

During my interviews and research of pattern seekers, I asked each interviewee to reflect on how patterns successfully inspired discovery. Based on their answers, I established that successful pattern seekers have the following characteristics in common. How do you match up?

1. Do you have an interest in and a mind for recognizing certain patterns? When did you first recognize you had this talent?

- Temple Grandin, who values autism as a gift, wondered that if we are differentiated by having different kinds of minds, why not acknowledge them as different and make them work to our advantage? Known as a visual thinker, Temple encourages people to recognize their thinking strengths and weaknesses and seek out ways to nurture their strengths.

- Growing up near the beaches of San Diego, Robert Ballard followed a different path than his peers. Instead of learning how to surf, Ballard, instead, took up scuba to explore patterns in nature below the surface. This helped launch his childhood dream of becoming an underwater explorer. Now known for his successful discovery of underwater mountain ranges, deep water vents, and sunken ships, Ballard credits search patterns in part to his success. Like Ballard, others have self-recognized their strengths and ambitions earlier in life.

- Truth wizard Renee Ellory had suspicions in her thirties that she was different. She had more than a casual interest in and talent for understanding human behavior patterns. That talent became a career of recognizing deception in someone and training others, such as Homeland Security and FBI personnel, also to be skilled in spotting deception.

- As a kid, Harvey College professor Arthur Benjamin enjoyed playing with numbers in his head and looking for patterns in numbers. This eventually led to his methods of doing fast mental math. Arthur Benjamin, "America's Best Math Whiz," according to Reader's Digest, knew by the eighth grade that he had a knack for solving math problems through using number patterns (Benjamin, 2015).

- "I was in my algebra class, and the teacher left a problem on the board as 108 squared. Being the hyperactive child that I was, I blurted out, well that's 11,664, and she said, that's right, how did you do that? And I said, oh, it was easy, I did 100 times 116, added 8 squared, and got 11,664. Well, she was very smart and she knew her algebra well enough to know what was going on, and why it would work. But she had not seen that method before, and no one at the middle school or high school had seen that, so it was my first experience with discovering something new, at least new to the people in my surroundings" (Benjamin, 2015).

- In data scientist Gilberto Titericz's opinion, math is a talent that you are born with, at least in part. "Of course, people can learn it from school, but many people, including me, feel inspired to use it to solve complex problems. For example, if someone asks me if I would like to play video game, play soccer, or solve a machine-learning problem I would answer: 'Tell me about that problem.' Solving very hard data problems are extremely fun to me. A good example is that I learned data science by myself just to find a way to optimize my investments in the Brazilian stock exchange."

2. Do patterns spark your curiosity? Do you have a passion to know what patterns can tell you?

- It has been suggested that the well-known pattern seeker Albert Einstein got his mathematical gifts from his father and his love of music from his mother, but he himself disagreed, stating curiosity, obsession, and sheer perseverance brought him to his ideas (Fitzgerald and O'Brien, 2007).

- Curiosity and a "passion to know" are often paired. It's not enough to be curious. Making a new discovery requires someone to follow up to interpret the meaning and significance of the pattern. Just a week after she had been looking at galaxies, Hanny van Arkel noted a startling green cluster on one image. A few others had looked at the same picture, presumably not knowing what to make of the cluster and moved on. Van Arkel didn't do that. "Some friends have suggested to me that the curiosity and the need for the answer is very much part of my personality."

- Polly Murray recognized many patterns about the disease she and her family suffered for many years. It was also shared by 39 adults and children living in the very same neighborhoods of Lyme, Connecticut; pain, redness, and rashes characteristic of rheumatoid arthritis. Her passion to know the cause and cure for the disease led her to contact expert after expert until she found Allen Steere, who discovered that an insect was the vector for the disease. The disease was coined Lyme disease, a disease now present in every state.

- As a child, Robert Ballard was always curious about things and he was fortunate enough not to have that passion extinguished as he grew up. "E/V Nautilus and the Ocean Exploration Trust gave me the opportunity to pour fuel on the flames of the public's curiosity to help keep it alive for them." His rage to know rubbed off on the thousands of youngsters who accompanied him virtually on his deep-water missions on the Nautilus.

- Environmental historian Dietrich Schlobohm often hiked the rural areas of western Massachusetts. When he saw stone walls, an alarm bell went off. The stone walls told Dietrich that at some earlier time humans lived here. He would immediately begin to look for the home site. In most cases this was simply a cellar hole lined by stone. There would be no longer wood structures. He would mark it on his topographical map to figure out which town he was in. Dietrich would go to the town hall and search for records and to find out what family lived there at what given time. His research doggedly pieced together a fascinating story of land use, change, and abandoned farm communities.

3. Do you subscribe to Winston Churchill's famous quote, "never, never, never give up!"

- Dr. Allen Steere credits persistence. "You often have to persist a long time when other people are questioning whether you're right or not. It's been that way just all the way along for 40 years. I think it takes a certain passion about wanting to pursue some particular line of thinking and follow it as far as you can. I'm also learning that doing that generally involves more than you can do in one generation. A part of what I need to do now is train the next generation."

- Missing-people finder Judy Moore never gives up. Even if the percentage of finding someone alive after 24 hours decreases, Moore will continue searching. It took her three years working with volunteers and on her weekends to finally discover the body of a missing New Hampshire child. "Never stop looking" is what drives Moore's success.

- Robert Ballard began to dream about exploring the Titanic in 1967. The dream was inspired by a speaker describing the history of the 1912 sinking of the Titanic and the challenge to ever find her at 12,000 feet under the north Atlantic. It was 1973 when Ballard first thought it was possible, when a submarine designed to withstand depths up to 12,000 feet was built. For the next 12 years, he tried unsuccessfully to raise interest and money to launch an expedition to discover her. Finally discovering the Titanic in 1985, he succeeded in making his dream come true.

4. Are you entrepreneurial and motivated to turn patterns into products or new ways of doing things?

- Clarence Birdseye's intense curiosity led him to discover why fish could recover from freezing at only certain times of the year. Equipped with the knowledge from his discovery, Birdseye created a method of flash-freezing fish for sale at local supermarkets. His entrepreneurial spirit led to the founding of Birdseye Foods and a whole new frozen food industry. Because Birdseye revolutionized the way we eat, a frozen food section in your local supermarket is a matter of course.

- Gordon Moore predicted in the early 1960s that computer chip power would increase and that chip cost would decrease each year. That finding still drives the development of today's newest technologies. The creation of the smart phone, tablets, and smarter, faster computers are products of his predictions.

- When California researchers discovered that peer pressure was better at motivating customers to conserve energy than saving money or saving the environment, Opower of Massachusetts used this information to reduce their customers' energy use. Each month, customers receive graphic information that compares their energy use to all their neighbors and their most efficient neighbors. The more efficient customers are rewarded with smiley faces. It's a win-win situation for both the company and consumer. Together, they conserve energy, money, and the environment.

- Gilberto Titericz competes for money by finding a way to best mine big data for patterns. He is also good at it. He is currently a grandmaster, ranked number 1 out of 166,031. For example, he accurately forecasted the total daily incoming solar energy throughout the year at 98 Oklahoma solar-farm sites. His forecast will help Oklahoma utility companies find the right balance of renewable and fossil fuels to create electricity for their customers. Errors in the forecast could lead to large expenses for the utility from excess fuel consumption or emergency purchases of electricity from neighboring utilities.

5. Do you have a deep understanding and knowledge of your field of endeavor?

- All the interviewees acknowledged that it was important to them to have a deep understanding and knowledge of their own field of endeavor.

6. Do you practice, practice, practice?

- The ability to recognize patterns increases with practice. Gilberto Titericz often treats data several different ways, sometimes by trial and error, until a clear pattern emerges. Clarence Birdseye made several observations about fish survival after freezing to confirm his hypothesis. Playing with numbers is the way that Arthur Benjamin practices honing his skill of mental math.

- "Our visual sense is highly tuned to spot patterns, and this natural ability can be trained to achieve even greater levels of acuity. Improvements result primarily from learning which of the many possible patterns in data are meaningful to your work and then also learning—through practice—to spot these patterns quickly, even in a dense forest of visual noise. As with all skills, practice in pattern recognition is required to move from average ability to true expertise" (Few, 2006).

BIBLIOGRAPHY

A Citizen Science Project Tracking the Honey Bee Parasite Apocephalus Borealis (2012). Retrieved from ZomBee Watch: ZomBee Watch is a project that asks volunteers to track the behavior of honey bees that have been infected with a parasite.

Aberlin, M. (2015, July 15). *Intelligence Gathering-Disease Eradication in the 21st Century*. Retrieved from *The Scientist*: *www.the-scientist.com/?articles.view/article-No/43308/title/Intelligence-Gathering*

Adam, J. 2009. *A Mathematical Nature Walk*. Princeton: Princeton University Press.

Adams, T. (2012, March 17). *The Guardian - The Observer*. Retrieved from Galaxy Zoo and the New Dawn of Citizen

Science: *www.guardian.co.uk/science/2012/mar/18/galaxy-zoo-crowdsourcing-citizen-scientist*s

Almanac of American Philanthropy. (2012) Who Gives Most to Charity? *www.philanthropyroundtable.org*

Alper, J. (2016, April 8). *Belichick on sustained success: We've tried to be an outlier*. Retrieved from NBC Sports: *http://profootballtalk.nbcsports.com/2016/04/08/belichick-on-sustained-success-weve-tried-to-be-an-outlier*

Anderson, Erika, (2013, January 7). 3 Things You Can Do to Think Like a Genius. *Forbes*.

Anything Left Handed. (n.d.). *Advantages of being left-handed*. *www.anythinglefthanded.co.uk/being-lh/lh-info/advantages.html*

Applegate, Z. (2013, April 26). Guy Stewart Callendar Global Warming Discovery Marked. BBC News, *www.bbc.com/news/uk-england-norfolk-22283372*

Arden, R. (2014, August 20). *Genes Influence Young Children's Human Figure Drawings and Their Association With Intelligence a Decade Later*. Retrieved from *Psychological Science: http://pss.sagepub.com/content/early/2014/08/19/0956797614540686.full*

Artble. (2016). *Georges Seurat*. Retrieved from Artble: *www.artble.com/artists/georges_seurat*

Atela, P. (2003, March). *Fibonacci numbers- The Golden Angle*. Retrieved from Phyllotaxis: *www.math.smith.edu/phyllo/About/fibogolden.html*

Aubrey, A. (2008, August 1). *For Smokers, Quitting May Be Contagious.* Retrieved from New England Public Radio: *www.npr.org/templates/story/story.php?storyId=90681873*

Badger, E. (2014, April 10). *Researchers Have Figured Out How to Map the Social Influence of Public Smoking.* Retrieved from *The Washington Post: www.washingtonpost.com/news/wonk/wp/2014/04/10/researchers-have-figured-out-how-to-map-the-social-influence-of-public-smoking*

Baff, J. R. (1999). *Climate Controversies; Perspectives on Recent Climate Science and Policy.* Waterville, Maine: Colby College.

Bainbridge, C. (2012). *How Do Children Learn Language?* From About Parenting: *http://giftedkids.about.com/od/gifted101/a/language_learning.htm*

Ballard, R. (1988). *Exploring the Titanic.* Toronto: Madison Press Books.

Ballard, R. (2016, August 7). Interview. (R. Barkman, Interviewer)

Ballard, R. (2016, August 2). *The Discovery of the Titanic.* (R. Barkman, Interviewer)

Barkman, R. (1978). The Use of Otolith Growth Rings to Age Young Atlantic Silversides, Menidia menidia. *Transactions of the American Fisheries Society*, 790–792.

Barkman, R. (1991). *Coaching Science Stars-Peptalk and Playbook.* Tucson, AZ: Zephyr Press.

Batchelor, S. (2016, November 5). The Living Building. (R. Barkman, Interviewer)

Bates, J. (1984). Daddy of the Cage Game. In J. B. Williams, *Basketball Was Born Here* (p. 24). Springfield, MA: Springfield College.

Bellano, A. (2019, May 28). *Kids & Family*. Retrieved from Patch: *https://patch.com/new-jersey/cinnaminson/math-emagician-arthur-benjamin-performs-rcbc-april-12*

Bells, N. (2011, August 11). *Teen Aidan Dwyer uses Fibonacci sequence to make solar energy breakthrough*. Retrieved from Yahoo News: *https://ca.news.yahoo.com/blogs/dailybrew*

Bend, D. (2014, July 17). Opower. (R. Barkman, Interviewer)

Benjamin, A. (2015, August 15). Math Patterns. (R. Barkman, Interviewer)

Benjamin, A. (2015, August 15). The Magic of Math. (R. Barkman, Interviewer)

Benjamin, A. (2015). *The Magic of Math: Solving for X and Figuring Out Why*. New York: Basic Books.

Benyus, J. (1997). *Biomimicry*. Ney York: William Morrow Company.

Berne, J. (2013). *On a Beam of Light: A Story of Albert Einstein*. San Francisco: Chronicle Books.

Biological Science Curriculum Study. (n.d.). *BSCS 5E Instructional Model*. Retrieved from Biological Science Curriculum Study (BSCS): *https://bscs.org/reports/the-bscs-5e-in-structional-model-origins-and-effectiveness/*

Birdseye, C. (1910–1916). Journals 14 volumes. *Field Journal Collection*. Amherst College Archives and Special Collections.

Blodgett, S. (2018, August 2). Noah Text. (R. Barkman, Interviewer)

Blumenstock, J. A. (2015). Predicting poverty and wealth from mobile phone metadata. *Science*, 350, 1073–1076.

Bohannon, J. (2009, April 20). *Gamers Unravel the Secret Life of Protein*. Retrieved from Wired: *www.wired.com/2009/04/ff-protein*

Bolduc, G. (2014, February 10). Director of Meat and Seafood. (R. Barkman, Interviewer)

Bor, D. (2012). *The Ravenous Brain*. New York: Basic Books.

Boroditsky, L. (2010, July 23). *Lost in Translation*. Retrieved from *Wall Street Journal: www.wsj.com/articles/SB10001424052748703467304575383131592767868*

Borwein, J. (2015, July 20). *Moore's Law is 50 years old but will it continue?* Retrieved from PHYS ORG: *http://phys.org/news/2015-07-law-years.html*

Boulder, B. (2011, January 10). *Bloggers' Word Choice Bares Their Personality Traits, Study by CU-Boulder Researcher Finds*. See more at: *www.colorado.edu/today/2011/01/10/bloggers-word-choice-bares-their-personality-traits-study-cu-boulder-researcher-finds*

Boyle, R. (2010, July 16). *Smart Visual Algorithm Lets Un-manned Drones Perform Autonomous Search and Rescue Operations.* Retrieved from Popular Science: *www.popsci.com/technology/article/2010-07/small-drones-could-replace-search-and-rescue-helos-making-hiker-rescue-faster-and-cheaper*

Braccini, P. (2012, July 25). *Be a Planet Hunter–Help NASA Find New Planets Using Kepler Data.* Retrieved from *Crowdsourcing.org. See: https://blog.planethunters.org/tag/crowdsourcing/*

Brain Works Project. (2015). *Emotional Coping Brain.* Retrieved from Brain Works project: *www.copingskills4kids.net/Emotional_Coping_Brain.html*

Branswell, Helen. (2021, May 19) How the COVID Pandemic ends: Scientists look to the past to see the future, STAT. *www.statnews.com/2021/05/19/how-the-covid-pandemic-ends-scientists-look-to-the-past-to-see-the-future*

Brooks, D. (2013, May 20). *What Our Words Tell Us.* Retrieved from *The New York Times,* The Opinion Page: *www.nytimes.com/2013/05/21/opinion/brooks-what-our-words-tell-us.html?_r=3&*

Brown, C. B. (2015, January 15). *The Great Defender.* Retrieved from Grantland: *http://grantland.com/features/new-england-patriots-bill-belichick-coaching-legacy-super-bowl-seattle-seahawks*

Brown, D. (2017, April 11). *MASS Audubon.* Retrieved from Recommendations For Planting Have Changed: *https://blogs.massaudubon.org/yourgreatoutdoors/recommendations-for-planting-have-changed*

Buck, J. (1984). *House-Tree-Person Technique.* Test Critiques: *https://killianphd.com/pdf/House-Tree-Person%20(H-T-P). pdf*

Byrd, D. A. (2010). *Discovering Speech, Words, and Mind.* West Sussex, United Kingdom: Wiley-Blackwell.

Cain, S. (2013, October 13). *How Your Facebook Posts Reflect Your Personality Style.* Retrieved from Susam Cain: *www. thepowerofintroverts.com/2013/10/24/how-your-face-book-posts-reflect-your-personality-style*

Carpenter, E. (2009, June 9). *USF Professor's Research is Lies.* From University of San Francisco: *www.usfca.edu/news/ usf-professors-research-lies*

Carson, R. (1962). *Silent Spring.* Boston: Houghton Mifflin.

Carter-Saltzman, L. (1980). Biological and sociocultural effects on handedness; Between biological and adoptive families. *Science,* 1263–65.

Cassi, J. (2002). I know what you'll do next summer. *New Scientist,* 29–32.

Centers for Disease Control and Prevention. (2011). *Cigarette Smoking--United States, 1965-2008. www.cdc.gov/ mmwr/preview/mmwrhtml/su6001a24.htm*

Cermele, J. (2013, March 6). Bass Bait Test: We Rank 10 Brand-New Bass Lures. *Field and Stream.*

Chen, K. (2013, April). *The Effect of Language on Economic Behavior: Evidence from Savings Rates, Health Behaviors, and Retirement Assets.* Retrieved from American Economic Review: *www.anderson.ucla.edu/faculty/keith.chen/ papers/LanguageWorkingPaper.pdf*

Cherbuin, N. and Brinkman, C. (2006). Hemispheric interactions are different in left-handed individuals. *Neuropsychology*, 700–707.

Cherry, K. (2014, September 3). *How Many Emotions Are There?* From *About.com* Psychology: *http://psychology. about.com/b/2013/09/04/how-many-emotions-are-there. htm*

Christakis, M. and N.A. Fowler. (2008, May 22). *The Collective Dynamics of Smoking in a Large Social Network.* Retrieved from *The New England Journal of Medicine: www.nejm.org/doi/full/10.1056/NEJMsa0706154#t=articleResults&sref=https://delicious.com/rbarkman/search/ disease*

Chronicle of Philanthropy. (2012, August 19). How States Stack Up in Genorosity. *www.philanthropy.com/article/ how-states-stack-up-in-generosity*

Clugston, M. (2002, July/August). *The Fast Trackers.* Retrieved from Canadian Geographic: *http://halifaxsar.ca/about*

Coca-Cola Company. (2012, January 12). *5 Things You Never Knew About Santa Claus and Coca-Cola. www.coca-colacompany.com/company/history/five-things-you-never-knew-about-santa-claus-and-coca-cola*

Cook, M. E. (1993, January 1). *So you want to be a controller?* Retrieved from Aircraft Owners and Pilots Association: *www.aopa.org/News-and-Video/All-News/1993/January/1/Screen-Test*

Corley, J. (n.d.). *How To Write A Joke*. From Jim Corley's Comedy Clinic: *www.standupcomedyclinic.com/254/how-to-write-a-joke-2*

Courage, K. H. (2012, November 9). *Allergies from Pollen Projected to Intensify with Climate Change*. From *Scientific American*: *http://blogs.scientificamerican.com/observations/2012/11/09/allergies-from-pollen-projected-to-intensify-with-climate-change/*

Courtland, R. (2008, August 5). Green mystery blob may reveal black hole's last meal. *New Scientist*.

Crace, J. (2003, October 20). *Reading Between the Letters*. Retrieved from The Guardian: *www.theguardian.com/education/2003/oct/21/research.schools*

Croft, T. (1978, July). Nighttime Images of the Earth from Space. *Scientific American*, pp. 68–79.

Cukier, V. M. S. (2013). *Big Data: A Revolution That Will Transform How We Live, Work, and Think*. Boston: Eamon Dolan/Houghton Mifflin Harcourt.

Daland, R. (2009, June). Word Segmentation, Word Recognition, and Word Learning: A Computational Model of First. *Dissertation for Doctor of Philosophy*. Evanston, Illinois, USA: Northwestern University.

Dallas, G. (n.d.). *What are Fractals and why should I care?* Retrieved from George Dallas: *https://georgemdallas.wordpress.com/2014/05/02/what-are-fractals-and-why-should-i-care*

D'Amato, G. L. (2006). Thunderstorm-asthma and pollen allergy. *Allergy*, 11–16.

D'Amour-Daley, C. (2014, February 7). Vice President Big Y Supermarkets. (R. Barkman, Interviewer)

David, T. (2015, February 23). *The Most Important Word You'll Ever Use*. Retrieved from Psychology Today: *www.psychologytoday.com/blog/the-magic-human-connection/201502/the-most-important-word-youll-ever-use*

Deaton, M. (2012). *Using Fibonacci Analysis to Predict Market Breakouts*. Retrieved from Option Alpha: *http://optionalpha.com/using-fibonacci-analysis-to-predict-market-breakouts-11445.html*

Del Giudice, G., Weinberger, B., Grubeck-Loebenstein, B. (2014). Vaccines for the elderly. *Gerontology*. 2015; 61(*3*): 203-210.

Devlin, K. (1994). *Mathematics- The Science of Patterns*. New York, New York: Scientific American Library.

Devlin, K. (1998). *Life by the Numbers*. New York, New York: John Wiley and Sons.

Dewar, G. (2014, December 23). *What kids' drawings can reveal about their lives at home*. Retrieved from BabyCenter: *http://blogs.babycenter.com/mom_stories/kids-drawings-reveal-clues-12232014-about-life-at-home*

DeYoung, D. (2015, March 26). *Paws to Shoes*.

Diep, F. (2013, April 21). *A Test That Quantifies Basic Language-Learning Ability*. From Popular Science:

www.popsci.com/article/science/test-quantifies-basic-lan-
guage-learning-ability?src=SOC

Dietert, J. (2015, July 1). *The Sum of Our Parts*. Retrieved from
The Scientist: *www.the-scientist.com/?articles.view/arti-
cleNo/43379/title/The-Sum-of-Our-Parts*

Dizik, A. (2013, December 13). *Waiting Tables at Top-Tier
Restaurants Is New Career Path for Foodies*. Retrieved
from *The Wall Street Journal*: *www.wsj.com/articles/SB10
00142405270230413730457929350943682222*

Doan, A. (2012, November 29). *BIOMIMETIC ARCHITEC-
TURE: Green Building in Zimbabwe Modeled After Termite
Mounds*. Retrieved from Inhabitat: *http://inhabitat.com/
building-modelled-on-termites-eastgate-centre-in-zimba-
bwe*

Doke, J. (2012). *Analysis of Search Incidents and Lost Person
Behavior in Yosmite National Park*. Lawrence: University of
Kansas.

Donadio, R. (2013, June 30). *When Italians Chat, Hands and
Fingers Do the Talking*. Retrieved from *The New York
Times Europe*: *www.nytimes.com/2013/07/01/world/eu-
rope/when-italians-chat-hands-and-fingers-do-the-talking.
html*

Donges, J. (2009, July 1). *What Your Choice of Words Says
about Your Personality*. Retrieved from *Scienific American
Mind and Brain*: *www.scientificamerican.com/article/you-
are-what-you-say/?page=1*

Dubner, S. J. (2016, March 9). *The No-Tipping Point*. Retrieved
from Freakonomics: *http://freakonomics.com*

Ducap, D. (2013). *5 Supermarket Seafood Secrets*. Retrieved from *About.com* Fish and Seafood Cooking: *http://fish-cooking.about.com/od/howtochoosefreshfish/a/5-Super-market-Seafood-Secrets.htm*

Durayappah, A. (2010, January 5). *What Science Has to Say About Genuine vs. Fake Smiles*. From *Psychology Today; Thriving 101: www.psychologytoday.com/blog/thriv-ing101/201001/what-science-has-say-about-genuine-vs-fake-smiles*

Dwyer, A. (2011). *The Secret of the Fibonacci Sequence in Trees*. Retrieved from American Museum of Natural History: *www.amnh.org/learn-teach/young-naturalist-awards/winning-essays2/2011-winning-essays/the-secret-of-the-fibonacci-sequence-in-trees*

Edlow, J. (2003). *Bull's Eye: Unraveling the Medical Mystery of Lyme Disease*. New Haven: Yale University Press.

Eickhoff, T. C. (2008, August). *Penicillin: An Accidental Discovery Changed the Course of Medicine. www.healio.com/news/endocrinology/20120325/penicil-lin-an-accidental-discovery-changed-the-course-of-medi-cine*

Eimas, P. (1985, January). The Perception of Speech in Early Infancy. *Scientific American*, pp. 17–23.

Einstein Thought Experiments. (2009, September 9). Retrieved from NOVA: *www.pbs.org/wgbh/nova/physics/einstein-thought-experiments.html*

Ekman P. (1990). The Duchenne Smile: Emotional Exprtession and Brain Physiology II. *Journal of Personality and Social Psychology*, 342–353.

Ekman, P. (2003). *Emotions Revealed*. New york, New York: St. Martins's Press.

Ekman, P. (2009). *Telling Lies: Clues to Deceit in the Marketplace, Politics, and Marriage*. New York: W.W. Norton.

Eldredge, N. (1999). *The Pattern of Evolution*. New York: W.H. Freeman and Company.

Elwood, E, (2013, January 16). *Record-Breaking Early Flowering in the Eastern United States*. From PLOS: *www.plosone.org/article/info%3Adoi%2F10.1371%2Fjournal.pone.0053788*

Elliott, R. N. (2012). *The Wave Principle*. Chicago Snowball Publishing.

Entrepreneur. (2006, November 26). *How to Raise Money for a Nonprofit*. Retrieved from Entrepreneur: *www.entrepreneur.com/article/171296*

Erard, M. (2014, April 17). *Secret Military Test, Coming Soon to Your Spanish Class*. From Nautilus-Ideas Cognitive Science: *http://nautil.us/issue/12/feedback/secret-military-test-coming-soon-to-your-spanish-class*

Ercolini, A. (2009, January). *The role of infections in autoimmune disease*. Retrieved from Clinical and Experimental Immunology: *www.ncbi.nlm.nih.gov/pmc/articles/PMC2665673*

ExpressMED. (2013, September 9). *ExpressMED Blog—Fall Allergies, Why They're Bad and Getting Worst, and What to Do.*

Eyes for Lies. (2005, March 14). *Microexpressions—Test Yourself.* From Eyes for Lies: *www.eyesforlies.com/blog/2005/03/microexpressions-test-yourself*

Fairchild Camera & Instrument Corporation. (2007). *1965—"Moore's Law" Predicts the Future of Integrated Circuits.* Retrieved from Computer History Museum.

Farmer, B. (1979, April 8). The Shy Millionaire. *Parade Magazine.*

Few, S. (2006, November 25). *Visual Pattern Recognition.* Retrieved from Cognos- Innovation Center for Performance Management: *www.perceptualedge.com/articles/Whitepapers/Visual_Pattern_Rec.pdf*

Finson, K. (1995). Development and Field Test of a Checklist for the Draw-a-Scientist Test. *School Science and Mathematics,* 195–205.

Finson, K. (2014, September 12). *Drawing a Scientist: What We Do and Do Not Know After Fifty Years of Drawings.* Retrieved from Researchgate: *www.researchgate.net/publication/229447487_Drawing_a_Scientist_What_We_Do_and_Do_Not_Know_After_Fifty_Years_of_Drawings*

Fisheries and Oceans Canada. (2013, January 23). *"Earstones" Reveal the Life of Fish.*

Fitday. (n.d.). *Purchasing Frozen Fish: A Healthy Option?* Retrieved from Fitday: *www.fitday.com/fitness-articles/*

nutrition/healthy-eating/purchasing-frozen-fish-a-healthy-option.html

Fitzgerald, M. (2004). *Is There a Link Between Autism in Men and Exceptional Ability?* New York: Brunner-Rutledge.

Fitzgerald, M. and O'Brien. (2007, June 1). *Genius Genes— How Asperger Talents Changed the World.* Shawnee Mission: Autism Asperger Publishing Company.

Fleming, A. (1929, June). On the Antobacterial Action of Cultures of a Penicillium, with Special Reference to Their Use in the Isolation of Influenzae. *British Journal of Experimental Pathology,* 226–236.

Fleming, A. (1945, December 11). *Nobel Lecture. www.nobelprize.org/nobel_prizes/medicine/laureates/1945/fleming-lecture.html*

Fractal Foundation, (n.d.) Explore Fractals. *https://fractalfoundation.org*

Frank, L. (2009, November 2). *How the Brain Reveals Why We Buy.* Retrieved from Scientific American: *www.scientificamerican.com/article/neuromarketing-brain*

Friedman, U. (2014, January 8). *What You Get When 30 People Draw a World Map From Memory.* Retrieved from The Atlantic: *www.theatlantic.com/international/archive/2014/01/what-you-get-when-30-people-draw-a-world-map-from-memory/282901*

Frost, R. (2013). What Predicts Successful Literacy Acquisition in a Second Language? *Psychological Science,* 1243–1252.

Frye, P. (2013, January 12). *42 Alien Planets Found By Planet Hunters Amateur Astronomers.*

Gaines, C. (2014, July 2). *Soccer Popularity Is On The Rise In The US, But English Football Is Benefiting More Than MLS.* Retrieved from Business Insider: *www.businessinsider. com/soccer-popularity-english-football-mls-2014-7*

Galaxy Zoo. (n.d.). Retrieved from Galaxy Zoo: *www.gal-axyzoo.org*

Galaxy Zoo. (2016). *Classify Galaxies.* Retrieved from Galaxy Zoo: *www.galaxyzoo.org*

Gardner, H. (1983). *Frames of Mind.* New York: Basic Books.

Garland, T. H. (1987). *Fascinating Fibonaccis: Mystery and Magic in Numbers.* Palo Alto, CA: Dale Seymour Publications.

Gaukrodger, S. (2008). *Temporal Design for Pattern Recognition in Spatial Awareness.* London: European Organization for the Ssafety of Air navigation.

GeekFeminism.org. (2010, June 23). *Scientists are "normal" people, some children discover.* Retrieved from *Geek Feminism.org: http://geekfeminism.org/2010/06/23/sci-entists-are-normal-people-some-children-discover*

Gentilucci, M. B. (2006). Speech and gesture share the same communication system. *Neuropsychologia,* 178–90.

Gernhardt, A. (2013, February 20). *This Is My Family":- Differences in Children's FamilyDrawings Across Cultures.* Retrieved from Journal of Cross-Cultural

Psychology: *http://jcc.sagepub.com/content/ear-ly/2013/02/20/0022022113478658*

Ginsberg, J. (2009). Detecting influenza epidemics using search engine query data. *Nature* , 1012–1014.

Gladwell, M. (2005). *Blink: The Power of Thinking Without Thinking.* New York and Boston: Little, Brown and Company.

Gladwell, M. (2008). *Outliers.* New York: Little, Brown and Company.

Goodsell, D. (2000, September). *Lysozyme.* From RCSBP-DP-101.

Grandin, T. A. and Panek, R. (2014). *The Autistic Brain: Helping Different Kinds of Minds Succeed.* Boston: Mariner Books.

Grandin, T. (n.d.). *Visual Thinking- My Experiences with Visual Thinking Sensory Problems and Communication Difficulties.* Retrieved from Autism Research Institute: *http://www.autism.com*

Granville, V. (2013, April 17). *The amateur data scientist and her projects.* Retrieved from Vincent Granville's Weekly Digest: *www.analyticbridge.com/profiles/blogs/the-amateur-data-scientist-and-her-projects*

Griskevicius, V. (2010, March). Going green to be seen: status, reputation, and conspicuous conservation. *J Pers Soc Psychol.*, 392–404.

Group, The Barber (2012, February 7) "Don't We All See the World Through Colored Filters?"

Halberstam, D. (2005). *The Education of a Coach.* New York: Hyperion.

Hardy, G. H. (1940). *A Mathematician's Apology.* Cambridge, UK: Cambridge University Press.

Hare, R. (1982). New Light on the History of Penicillin. *Medical History,* 1–24.

Hartshorne, J. (2014, June 2). *Which English?* Retrieved from Games with Words: *http://gameswithwords.org*

Hauser, A. (2015). *Allergies Worse In or After Rain, Allergists Say.* From The Weather Channel—Health: *https://weather.com/health/allergy/news/allergies-worse-or-after-rain-allergists-say-20130912*

Hensley, N. (2016, September 16). *Homeless man who found bag of bombs near New Jersey train station wanted backpack for job search.* Retrieved from *The New York Daily News: www.nydailynews.com/news/national/good-samaritans-recall-finding-bag-bombs-new-jersey-article-1.2800236*

Hill, D. O. (2007). *Managing the Lost Person Incident.* National Association for Search and Rescue.

Holdrege, C. (2000, Fall). *In Context- The Skunk Cabbage.* From The Nature Institute: *www.natureinstitute.org/pub/ic/ic4/skunkcabbage.htm*

Horowitz, A. (2013). *On Looking—A Walker's Guide to the Art of Observation.* New York: Scribner.

Hough, M. T. (2014, January 13). *VerbCornerA Window Into the Brain One Thought at a Time*. Retrieved from SciStarter: *http://scistarter.com/blog/tag/linguistics/#sthash. Uo34zy61.zRnNmbm4.dpbs*

How to Analyze a Child's Drawings. (2013, June 19). Retrieved from Hub Pages: *http://hubpages.com/education/How-to-Analyze-a-Childs-Drawings*

Hughes, P. (2001, February 8). *On the Shoulders of Giants: Alfred Wegener.* Retrieved from NASA Earth Observatory: *http://earthobservatory.nasa.gov/Features/Wegener/wegener.php*

Hurley, D. (2012, April 12). *Can You Make Yourself Smarter?.* Retrieved from *The New York Times Magazine: www.nytimes.com/2012/04/22/magazine/can-you-make-yourself-smarter.html?pagewanted=all&_r=0*

Hutcheson, G. D. (2005, Spring). Moore's Law: The History and Economics of an Observation that Changed the World. *The Electrochemical Society Interface*, pp. 17–21.

iAWAKE. (2016, January 6). *Fractal Entertainment: A New Psychoacoustic Technology Inspired by Nature.* Retrieved from iAwake: *www.iawaketechnologies.com/fractal-entrainment-new-psychoacoustic-tech-inspired-nature*

In the News. (2014, February). *Online World War I Diaries Tell of Soldiers' Daily Lives.* Retrieved from In the News: *http://hmhinthenews.com/online-world-war-i-diaries-tell-about-soldiers-daily-lives*

I Q Test Labs (2015). https://intelligencetest.com

Jacobson, M. (2015, January 25). *The Rise and Fall (and Rise) of the Ukulele.* Retrieved from *The Atlantic:* *www.theatlantic.com/entertainment/archive/2015/01/though-it-be-little-the-rise-of-the-ukulele/384453*

Jaffe, E. (2010, December). *The Psychological Study of Smiling.* From Association of Psychological Science; Observer: *www.psychologicalscience.org/index.php/publications/observer/2010/december-10/the-psychological-study-of-smiling.html*

Jalan, S. (2014, March 7). *Top 10 Most Loved Sports in the World.* From List Dose; Just Enough to be Addictive.

Janson, T. (2012). *Thye History of Langauges.* Oxford: Oxford University Press.

Jarboe, E. C. (2002, September). *Art Therapy: A Proposal for Inclusion in School Settings.*

Jenkins, A. (n.d.). *Climate change: How do we know?* From National Aeronautics and Space Administration: *http://climate.nasa.gov/evidence*

Johnson, S. (2006). *The Ghost Map.* New York: Riverhead Books.

Jonas, G. (2014, April 16). *Science's Amateur Hour.* Retrieved from Newsweek: *www.newsweek.com/2014/04/25/sciences-amateur-hour-248155.html*

Jones, C. (2015, March 31). *Millions of Americans have little to no money saved.* Retrieved from USA Today: *www.usatoday.com/story/money/personalfinance/2015/03/31/millions-of-americans-have-no-money-saved/70680904*

Jones, K. A. (1992). *Food search behaviour in fish and the use of chemical lures in commercial and sports fishing.* London: Springer.

Judson, H. F. (1980). *The Search for Solutions.* New York: Holt Rinehart and Winston.

Jurvetson, S. (2015). "Transcending Moore's Law to forge the future" CORE, pp. 38-40.

Kageyama, C. J. (1999). *What Fish See: Understanding Optics and Color Shifts When Designing Flies.* Portland: F. Amato.

Kahn, B. (2013, September 16). *Human Fingerprints Visible in Atmospheric Changes.* From Climate Central: *www.climatecentral.org/news/human-fingerprints-visible-in-atmospheric-changes-16482*

Kapeleris, J. (2010, August 6). *The Power of Creative Visualization.*

Karch, M. (n.d.). *Google Books Ngram Viewer.* Retrieved from About Tec: *http://google.about.com/od/n/a/Google-Books-Ngram-Viewer.htm*

Kay, Alan, WikiQuote. *https://en.wikiquote.org/wiki/Alan_Kay*

Keim, B. (2010, August 10). *Minds Beat Machines in Protein Puzzle Showdown.* Retrieved from *Wired: www.wired.com/2010/08/crowdsourced-protein-folding*

Kidd Celeste, K. S. (2011). Toddlers use speech disfluencies to predict speakers' referential. *Developmental Science*, 925–934.

Kids Discover. (2016). *World War I*. Retrieved from Kids Discover: *www.kidsdiscover.com/shop/issues/world-war-i-for-kids*

Kilgore, T. (2015, June 16). *5 charts to help unravel the Elliott Wave mystery*. Retrieved from Market Watch: *www.marketwatch.com/story/5-charts-to-help-unravel-the-elliott-wave-mystery-2015-06-08*

Kitayama, S. D. (2008, September). *Perceiving an object in its context—is the context cultural or perceptual?* Retrieved from Journal of Vision: *http://jov.arvojournals.org/article.aspx?articleid=2193055*

Klass, P. (2011, March 6). *On the Left Hand, There Are No Easy Answers*. Retrieved from *The New York Times; View: www.nytimes.com/2011/03/08/health/views/08klass.html?_r=1*

Koba, M. (2014, July 24). *Sports Business Golf is on the decline in the US because...* Retrieved from CNBC: *www.cnbc.com/id/101860445*

Koch, W. (2013, May 31). *Climate change linked to more pollen, allergies, asthma*. From USA Today: *www.usatoday.com/story/news/nation/2013/05/30/climate-change-allergies-asthma/2163893*

Krulwich, R. (2009, December 19). *There's A Fly In My Urinal*. Retrieved from NPR News: *www.npr.org/templates/story/story.php?storyId=121310977*

Kurlansky, M. (2012). *Birdseye—The Adventures of a Curious Man*. New York: Random Houser.

Kurzweil, R. (2012). *How to Create a Mind: The Secret of Human Thought Revealed.* New York: Penguin Books.

Kyed, D. (2016, April 8). *Bill Belichick: Being 'Outlier' Has Led To Patriots' Sustained Success.* Retrieved from NESN: *http://nesn.com/2016/04/bill-belichick-being-outlier-has-led-to-patriots-sustained-success*

Lanny Lin, M. R. (2010, January). Supporting Wilderness Search and Rescue Efforts with Integrated Intelligence: Autonomy and Information at the Right Time and the Right Place. *Association or the Advancement of Artificial Intelligence,* 1542–1547.

Lasky, Alex, (2013) TED Talks. How behavior science can lower your energy bill. *www.ted.com/speakers/alex_laskey*

Leggett, S. (2014, January 14). *Operation War Diary—Your Archive Needs You!.* Retrieved from The National Archives: *http://blog.nationalarchives.gov.uk/blog/operation-war-diary-archive-needs*

Lewis, D. (2013). *The Brain Sell- When Science Meets Shopping.* London: Nicholas Brealy Publishing.

Li, M. (2013, May 29). *Keeping Up with Moore's Law.* Retrieved from Dartmouth Undergraduate Journal of Science: *http://dujs.dartmouth.edu*

Lindholm, S. (2014, February 10). *Major League attendance Trends Past, Present, and Future.* Retrieved from SB Nation: *www.beyondtheboxscore.com/2014/2/10/5390172/major-league-attendance-trends-1950–2013*

Long, T. (2009, May 15). *May 19, 1910: Halley's Comet Brush-es Earth With Its Tail.* Retrieved from *Wired: www.wired. com/2009/05/dayintech-0519*

Love, J. (1986). *McDonald's: Behind the Arches.* New York: Bantam.

Malchiodi, C. (2009, July 15). *Helping Children Draw Out Their Traumas.* Retrieved from *Psychology Today: www. psychologytoday.com/blog/arts-and-health/200907/help-ing-children-draw-out-their-traumas*

Mandelbrot, Benoit, (1983). The Fractal Geometry of Nature. W.H. Freeman.

Markel, H. (2013, 27 September). *The Real Story Behind Peni-cillin.* From PBS Newshour: *www.pbs.org/newshour/run-down/the-real-story-behind-the-worlds-first-antibiotic*

Mason, C. J. (1991, May–June). Draw-a-Scientist Test: Future Implications. *School Science and Mathematics*, 91–95.

Masoumi, S. (2012). *Phyllotactic Tower Proto-type Mimics Plants.* Retrieved from Solaripedia: *www.solaripedia.com/13/404/6197/structure_of_cone.html*

Masthoff, G. R. (2011, August 24). *Welcome to the Joking Computer.* From The Joking Computer: *www.abdn.ac.uk/ncs/departments/computing-science/joking-comput-er-309.php*

Mayer-Schonberger, V. (2013). *Big Data.* New York: Houghton Mifflin Harcourt.

Maynard, M. (2007, July 4). *Say 'Hybrid' and Many People Will Hear 'Prius'*. Retrieved from *The New York Times Business Day*: *www.nytimes.com/2007/07/04/business/04hybrid.html*

Macarthur, Robert H. (1984) Geographical Ecology: Patterns in the Distribution of Species

McCarthy, C. (2008, October 19). *Can you read this?* Retrieved from Learn English- A Lesson A Day: *www.ecenglish.com/learnenglish/lessons/can-you-read*

McConnell, B. B. (2013, May 3). *Rabble.ca*. Retrieved from Big Data: The role of citizen scientists in the age of information abundance: *http://rabble.ca/news/2013/05/big-data-role-citizen-scientists-age-information-abundance-0*

McElhinney, P. (2013, February 20). *How Tennis Has Changed Over The Last 30 Years*. Retrieved from Steve Tennis: *www.stevegtennis.com/2013/02/how-tennis-has-changed-over-the-last-30-years*

McFarlane, G. (1984, March 1). Alexander Fleming in Fact and Fantasy. *New Scientist*, pp. 26–28.

McKay, F. (1929, July). The establishment of a definite relation between enamel that is defective in its structure, as mottled enamel, and the liability to decay. *The Dental Cosmos*, 747–755.

McManus, I. (1999). Handedness, cerebral lateralization and the evolution of language. In M. C. Lea, *The descent of mind: Psychological perspectives on hominid evolution* (pp. 194–217). Oxford University Press.

Middlebrook, J. (2015, June 15). Relationship Among Plants, Architecture and Numbers. (R. Barkman, Interviewer)

Miemis, V. (2010, April 10). *Essential Skills for 21st Century Survival: Part I: Pattern Recognition.* Retrieved from Emergent by Design: *http://emergentbydesign.com/2010/04/05/essential-skills-for-21st-century-survival-part-i-pattern-recognition*

Miko, I. (2008). *Gregor Mendel and the Principles of Inheritance.* Retrieved from SciTable by Nature Educaion: *www.nature.com/scitable/topicpage/gregor-mendel-and-the-principles-of-inheritance-593*

Miller, P. (2012, January). *A Thing or Two About Twins.* National Geographic. *www.nationalgeographic.com/magazine/article/identical-twins-science-dna-portraits*

Miller-Rushing, R. P. (2012, February). Uncovering, Collecting, and Analyzing Records to Investigate the Ecological Impacts of Climate Change: A Template from Thoreau's Concord. *Bioscience,* 170–181.

Mitchell, J. (2005). *Rapala: Legendary Lures.* Minneapolis: Voyageur Press.

Mohammed, M. (2003, January 29). *Vaux Conducts Survey for Online Dialect Atlas.* Retrieved from *The Harvard Crimson: http://dialect.redlog.net*

Moore, G. (2013, February 27). Computer on Chip and Moores Law. (R. Barkman, Interviewer)

Moore, J. (2014, January 28). Chair of Emergency Medicine Dept. (R. Barkman, Interviewer)

Moore, M. (2012, September 12). *Science in Our World: Certainity and Controversy*. Retrieved from Earworms are Taking Over Your Brain: *www.personal.psu.edu/afr3/blogs/siowfa12/2012/09/earworms-are-taking-over-your-brain.html*

Morris, C. M. (2010, June 3). Citizens-as-data-analysts benefit... all of us. *The Chronicle of Higher Education.*

Mother Nature Network. (2010, February 10). *7 Amazing Examples of Biomimicry*. From Mother Nature Network: *www.mnn.com/earth-matters/wilderness-resources/photos/7-amazing-examples-of-biomimicry/copying-mother-nature*

Mottron, L. (2011, November 3). The Power of Autism. *Nature, 479*, 33–34.

Mountain Rescue Association. (2011, August 11). *What to Do If You Get Lost*. Retrieved from Mountain Rescue Association: *http://mtrescueassoc.blogspot.com/2011/08/what-to-do-you-get-lost-imagine.html*

Murray, P. (1996). *The Widening Circle—A Lyme Disease Pioneer Tells Her Story*. New York: St. Martin's Press.

Naismith, J. (1891). *Rules for Basketball*. Springfield, MA: The Grimger Publishing Company.

Naismith, J. (1941). *Basketball Origins—Creative Problem Solving in the Guilded Age*. New York: Association Press.

Narayan, N. (2011, September 28). *Brand Management Study at Amazon*. Retrieved from Slide Share: *www.slideshare.net/nik123hil/brand-management-study-of-amazon*

NASA Ames Research Center. (2013, March 13). *In the Zone: How Scientists Search for Habitable Planets.* Retrieved from NASA Ames Research Center: *https://www.nasa. gov/mission_pages/kepler/news/kepler20130717.html*

National Academy of Science. (1999). *How People Learn.* Washington DC: National Academy Press.

National Aeronautics and Space Administration. (2000, October 23). *Visible Earth.* Retrieved from National Aeronautics Space Association: *http://visibleearth.nasa.gov/view. php?id=55167*

National Human Genome Research Institute. (2012, June 13). *National Human Genome Research Institute.* Retrieved from *Genome.gov: www.genome.gov/25520880*

National Institute of Alcohol Abuse and Alcoholism. (2012, June). *A Family History of Alcoholism.* From National Institue of Alcohol Abuse and Alcoholism: *https://pubs. niaaa.nih.gov*

National Institutes of Health. (2011, March 25). *National Institute of Dental and Craniofacial Research.* Retrieved from The Story of Fluoridation: *www.nidcr.nih.gov/oralhealth/ topics/fluoride/thestoryoffluoridation.htm*

National Research Council. (2005.) *How Students Learn: History, Mathematics, and Science in the Classroom.* Washington, DC: The National Academies Press.

National Research Council. (2012.) *A Framework for K-12 Science Education: Practices, Crosscutting Concepts, and Core Ideas.* Washington, DC: The National Academies Press.

National Wildlife Federation. (n.d.). *Allergies and Global Warming*. From National Wildlife Federation: *www.nwf.org/Wildlife/Threats-to-Wildlife/Global-Warming/Global-Warming-is-Causing-Extreme-Weather/Allergies.aspx*

Nerhardt, G. (1970, November). Humor and inclination to laugh:emotional reactions to stimuli of different divergence from a range of expectancy. *Scandinavian Journal of Psychology*, 185–195.

Newson, L. (n.d.). *Absolute Risk and Relative Risk*. Retrieved from Patient: *http://patient.info/health/absolute-risk-and-relative-risk*

Newton, Isaac, "Letter from Sir Isaac Newton to Robert Hooke." Historical Society of Pennsylvania. Retrieved 7 June 2018.

Nguyen, T. (2011, August 22). *Why 13-year-old's solar power 'breakthrough' won't work*. Retrieved from ZD Net: *www.zdnet.com/article/why-13-year-olds-solar-power-breakthrough-wont-work*

Nidetch, J. (2009). *The Jean Nidetch Story: An Autobiography*. Weightwatchers.

Nobelprize.org. (1945, December 11). *Nobel Lecture- Penicillin*. From *nobelprize.org*: *www.nobelprize.org/nobel_prizes/medicine/laureates/1945/fleming-lecture.html*

Nobelprize.org. (1945). *Sir Howard Florey—Biographical*. From *Nobelprize.org*: *www.nobelprize.org/nobel_prizes/medicine/laureates/1945/florey-bio.html*

Nobelprize.org. (1945, December 10). *The Nobel Prize in Physiology or Medicine 1945 (Banquet Speech).* From *www.nobelprize.org/prizes/medicine/1945/fleming/speech*

Nobelprize.org. (n.d.). *The Discovery of the Molecular Structure of DNA - The Double Helix.* Retrieved from *Nobelprize.org: www.nobelprize.org/educational/medicine/dna_double_helix/readmore.html*

Noë, A. (2014, May 7). *What Is The Funniest Joke In The World?* From National Public Radio-Cosmos and Culture: *www.npr.org/blogs/13.7/2014/03/07/287250640/what-is-the-funniest-joke-in-the-world*

O'Connor, K. H. (2007). *Managing the Lost Person Incident.* Centreville, VA: National Association for Research and Rescue.

Oech, R. V. (2008). *A Whack on the Side of the Head, 25th Anniversary Edition Revised.* New York: Grand Central Publishing.

O'Sullivan, M. (2009, March 23). *Deception; The Truth About Truth Telling.* From *Psychology Today: www.psychologytoday.com/blog/deception*

Panaggio, D. M. (2012). A model balancing cooperation and competition can explain our right-handed world. *Journal of the Royal Society Soc. Interface, https://royalsocietypublishing.org*

Pandit, S. (2013, April 22). *What Does Your Walk Reveal About Your Personality?*

Papadatou-Pastou, M. (2011, August). Handedness and language lateralization: Why are we right handed and left-Brained? *Hellenic Journal of Psychology*, 248–265.

Parmar, R. (2014, January–February). *The New Patterns of Innovation*. Retrieved from Harvard Business Review: *https://hbr.org/2014/01/the-new-patterns-of-innovation*

Parry, W. (2012, March 8). *Thoreau's Notes Reveal How Spring Has Changed in 150 Years*. From *Live Science: www.livescience.com/18938-thoreau-citizen-science-climate-change.html*

PBS. (1998). *Fleming discovers penicillin*. From A Science Odyssey- People and Discoveries: *www.pbs.org/wgbh/aso/databank/entries/dm28pe.html*

PC Plus. (2011, August 7). *How to explore space from your desktop*. Retrieved from PC Plus: *www.techradar.com/news/software/applications/how-to-explore-space-from-your-desktop-987555*

Pennebaker, J. (2011). *The Secret Life of Pronouns*. New York: Bloomsbury Press.

Pierce, J. (2013, January 13). *The Words You Use On Your Resume Reflect Your Personality Traits*.

Pincott, J. (2012, March 13). *Slips of the Tongue*. Retrieved from *Psychology Today: www.psychologytoday.com/articles/201203/slips-the-tongue*

Pinker, S. (2005, July). *Steven Pinker: What our language habits reveal*. Retrieved from TED Talks: *https://www.ted.*

com/talks/steven_pinker_what_our_language_habits_reveal?language=en

Pro Football Hall of Fame. (2013, October 13). *Pro Football Hall of Fame—Just like Dad!* From Pro Football Hall of Fame: *www.profootballhof.com/history/stats/fathers.aspx*

Proust, Marcel. La Prisonnière', *Remembrance of Things Past* (also known as *In Search of Lost Time*), Volume 5.

Reading Rockets. (2011). *Helping Stuggling Readers.* From Reading Rockets: *www.readingrockets.org/article/patterns-and-categorizing*

Reed, S. B. (2007). *A Tale of Two Brains: How REMAP Can Help.* Retrieved from The Remap Process: *https://psychotherapy-center.com/therapy-methods/remap/introducing-quick-remap/a-tale-of-two-brains-how-remap-can-help*

Reinecke, K. (2016). *Are You More Eastern or Western?* Retrieved from Lab in the Wild: *www.labinthewild.org/studies/frame-line*

Renee. (2014, February 12). Author of Eyes for Lies website. (R. Barkman, Interviewer)

Rian, I. M. (2014, September 23). *Tree-inspired dendriforms and fractal-like branching structures in architecture: A brief historical overview.* Retrieved from Researchgate: *www.researchgate.net/publication/264972592_Tree-inspired_dendriforms_and_fractal-like_branching_structures_in_architecture_A_brief_historical_overview*

Ritchie, G. A. (2011, August 24). *Humour and Communication.* From The Joking Computer: *www.abdn.ac.uk/ncs/departments/computing-science/joking-computer-309.php*

Roach, J. (2016, August 27). *Girls in Science Camp Reflection – Draw a Scientist Revelation.* Retrieved from Engage: *http://engageduniversity.blogs.wesleyan.edu/2015/08/27/girls-in-science-camp-reflection-draw-a-scientist-revelation*

Robinson, S. (Director). (2011). *Do You See What I See?* [Motion Picture].

Roland, C. (2006). *Young in Art.* Retrieved from artjunction: *www.artjunction.org/young_in_art.pdf*

Rooney, B. (2007, April 4). *UPS Figures Out the 'Right Way' to Save Money, Time and Gas.* Retrieved from ABC News: *http://abcnews.go.com/WNT/story?id=3005890&page=1*

Roush, W. (2009, March 16). Weaving Words with Wordle: A Talk with IBM's Jonathan Feinberg. Xconomy.

Rosales, G. (2014, January 5). *NM: Four lost hikers found in forest.*

Saffran, J. A. (1996, December). Statistical Learning by 8-Month-Old Infants. *Science,* 1926–1928.

Saleh Masoumi/Phyllotaxy Towers. (2014, March 17). Retrieved from Wikipedia: *http://en.wikipedia.org/wiki/User:Saleh_Masoumi/Phyllotaxy_towers*

Schara, R. (n.d.). *Our Story.* From Rapala: *www.rapala.com/content/rapala-general-information/our-history.html*

Schlobaum, D. (2014, May 15). The Mystery of Cellar Holes. (R. Barkman, Interviewer)

Schultz, C. K. (2012, July). Use of a lure in visual census significantly improves probability of detecting predatory fish. *Fisheries Research*, 70–77.

ScienceDaily (2011, March 15). *Bilinguals see the world in a different way, study suggests.* Retrieved from *ScienceDaily: www.sciencedaily.com/releases/2011/03/110314132531.htm*

ScienceDaily (2013, May 28). *Picking up a second language is predicted by ability to learn patterns.* From *ScienceDaily: www.sciencedaily.com/releases/2013/05/130528143800.htm*

ScienceDaily (2014, April 3). *Unbreakable' security codes inspired by nature.* From *ScienceDaily: www.sciencedaily.com/releases/2014/04/140403132111.htm*

ScienceDaily: *Science News.* (2012, April 25). *Shedding Light on Southpaws: Sports data help confirm theory explaining left-handed minority in general population. www.sciencedaily.com/releases/2012/04/120425140457.htm*

Seinfeld, J. (2012, December 12). *Jerry Seinfeld Interview: How to Write a Joke | The New York Times.* From You-Tube: *www.youtube.com/watch?v=itWxXyCfW5s*

SETI. (2016). Retrieved from SETI Home: *http://setiathome.berkeley.edu*

Shah, P. (1997). *A Model of the Cognitive and Perceptual Processes in Graphical Display Comprehension.* Palo Alto: AAAI Technical Report FS-97-0.

Shapiro, A. (2013, January 23). *From The Trenches To The Web: British WWI Diaries Digitized.* Retrieved from National Public Radio: *www.npr.org/sections/parallels/2014/01/23/264532419/from-the-trenches-to-the-web-british-wwi-diaries-digitized*

Shillum, M. (2011, July). *Brands as Patterns.* Retrieved from Shomi Partnership/Rogers Communication: *https://medium.com/method-perspectives/brands-as-patterns-b265d10ee7c7*

Shontell, A. (2011, March 24). *Why UPS Is So Efficient: "Our Trucks Never Turn Left".* Retrieved from Business Insider: *www.businessinsider.com/ups-efficiency-secret-our-trucks-never-turn-left-2011-3*

Sidwell, J. (2013, May 30). *The rise and rise of the ukulele!* Retrieved from Musicradar: *www.musicradar.com/guitart-echniques/the-rise-and-rise-of-the-ukulele-246654*

Silverman, S. (2014, January 16). *Why Is the Game of Basketball So Popular?* From Lifestrong.Com: *www.livestrong.com/article/364098-why-is-the-game-of-basketball-so-popular*

Simmon, R. (2008, April 22). *Cities at Night: The View from Space.* Retrieved from Earth Observatory: *http://earthobservatory.nasa.gov/Features/CitiesAtNight*

Simons, D. (2010, April 28). The Monkey Business Illusion. Retrieved from *www.youtube.com/watch?v=IGQmdoK_ZfY*

Simons, D. (2012, September). *But Did You See the Gorilla? The Problem With Inattentional Blindness*. Retrieved from *Smithsonian Magazine: www.smithsonianmag.com/science-nature/but-did-you-see-the-gorilla-the-problem-with-inattentional-blindness-17339778/?no-ist*

Simons, D. A. (2010). *The Invisible Gorilla*. Retrieved from The Invisible Gorilla: *www.theinvisiblegorilla.com/gorilla_experiment.html*

Sinclair, A.R.E. and P. Arcese, A. S. (1995). Population Consequences of Predation-Sensitive Foraging. *Ecology*, 882–891.

Smist, J. and R. Barkman (1996, October 25). Self-Efficacy of Pattern Recognition in Science of Middle School Students. *Northeastern Educational Research Association*, (pp. 1-8). Ellenville, New York.

Smith, R. (2012, April 5). *Amateur Science and the Rise of Big Science*.

Spiegel, A. (2014, September 1). *Our Use Of Little Words Can, Uh, Reveal Hidden Interests*. Retrieved from Health News From NPR: *www.npr.org/blogs/health/2014/09/01/344043763/our-use-of-little-words-can-uh-reveal-hidden-interests*

Springfield Union. (1930, March 6). For the First Time Anywhere! The Most Revolutionary Idea in the History of Foods Will Be Revealed in Springfield Today. *Springfield Union*.

Stahl, L. (2012, July 15). *Temple Grandin: Understanding autism*. Retrieved from 60 Minutes Overtime: *www.cbsnews.com/news/temple-grandin-understanding-autism*

Stankovski, T. (2014). Coupling Functions Enable Secure Communications. *Physical Review X*, 1103.

Statistic Brain. (n.d.). *Most Played Sports Worldwide*. From Statistic Brain; Percentages, Numbers, Financials, Rankings: *www.statisticbrain.com/most-played-sports-worldwide*

Staudt, A. (n.d.). *Allergies and Global Warming*. From National Wildlife Federation: *www.nwf.org/Wildlife/Threats-to-Wildlife/Global-Warming/Global-Warming-is-Causing-Extreme-Weather/Allergies.aspx*

Steere, A. (2016, March 15). Lyme Disease. (R. Barkman, Interviewer)

Stengel, J. (2013, December 12). *10 Things Every Brand Can Learn From Coke*. Retrieved from *Forbes*: *www.forbes.com/sites/jimstengel/2013/12/12/10-things-every-brand-can-learn-from-coke*

Sternberg, R. J. (2008, May 13). Increasing fluid intelligence is possible after all. *Proc Natl Acad Sci U S A*, pp. 6791–6792.

Stevenson, D. K. (1992, January). Otolith microstructure examination and analysis. *Can. Spec. Pub.Fish. Aquatic. Sci.*, 117.

Stone, B. (2013). *The Everything Store*. New York: Little Brown and Company.

Suddath, C. (2010, June 15). *A Brief History Of Velcro.* From TIME: *http://content.time.com/time/nation/article/0,8599,1996883,00.html*

Sullivan, W. (1991). *Earth at Night: An Image of the Earth at Night Based on Cloud Free Satellite Images.* Retrieved from SAO/NASA Astrophysics Data System (ADS): *http://articles.adsabs.harvard.edu/cgi-bin/nph-iarticle_query?bibcode=1991ASPC...17...11S&db_key=AST&page_ind=0&plate_select=NO&data_type=GIF&type=SCREEN_GIF&classic=YES*

Swaminathan, N. (2008, April 24). *For the Brain, Cash Is Good, Status Is Better.* Retrieved from Scientific American: *www.scientificamerican.com/article/for-the-brain-status-is-better*

Symington, D. (2006, July 7). *The 'Draw a Scientist Test': Interpreting the Data.* Retrieved from Research in Science and Technological Education: *www.tandfonline.com/doi/abs/10.1080/0263514900080107?journalCode=crst20&*

Szeliga, B. (2011, September). *Is Frozen Fish Better Than Fresh Fish?*

Szent-Gyorgyi. (1985, October 16–18). *Bridging the present and the future: IEEE Professional Communication Society conference record, Williamsburg, Virginia,* p. 14.

Taub, D. (2010). *Effects of Rising Atmospheric Concentrations of Carbon Dioxide on Plants.* From The Nature Education Knowledge Project: *www.nature.com/scitable/knowledge/library/effects-of-rising-atmospheric-concentrations-of-carbon-13254108*

Tennant, B. (2012, May 2). *How To Use These 3 Hypnotic "Power Words" To Covertly Increase Your Conversion Rates*. Retrieved from Kissmetrics: *https://blog.kissmetrics. com/3-hypnotic-power-words*

Thayer, J. (2000, September). Sex Differences in Judgement of Facial Affect: A Multivariate Analysis of Recognition Errors. *Scandinavian Journal of Psychology*, 243–246.

The Any Day Gourmet. (2013). *Flash Frozen vs. Fresh*.

The Behavioural Insights Team. (2012, July 4). *Behavioural Economics-Developing solutions for better decision making*. Retrieved from INudgeYou: *www.inudgeyou.com*

The Cornell Laboratory of Ornithology. (2011). *Yard Map*. Retrieved from The Cornell Laboratory of Ornithology: *http://content.yardmap.org*

The Human Camera. (2013, December 13). Retrieved from YouTube: *https://www.youtube.com/watch?v=phkNg-C8Vxj4*

The Juniper Company. (2013). *Your brand is what people say about you when you're not in the room"*. Retrieved from The Juniper Company: *www.thejuniperco.co.uk/ news/2013/january/your-brand-what-people-say-about-you-when-youre-not-room-jef*

The Nature Conservancy. (n.d.). *Journey with Nature--The Skunk Cabbage*.

The Planetary Society. (2016). *SETI@home*. Retrieved from The Planetary Society: *www.planetary.org/explore/projects/seti/seti-at-home.html*

The Top Tens. (2013). *Top Ten Greatest Sports*. From The Top Tens: *www.thetoptens.com/top-ten-greatest-sports*

Thomas, L. (1987). *The Search for Solutions*. Baltimore: The John Hopkins Press.

Thompson, C. (2003, October 26). *There's a Sucker Born in Every Medial Prefrontal Cortex*. Retrieved from *The New York Times Magazine: www.nytimes.com/2003/10/26/magazine/there-s-a-sucker-born-in-every-medial-prefrontal-cortex.html?pagewanted=1*

Thompson, C. (2014, March 14). *Charles Viancin Silicone Lids: Do they work?* From *Komo News.com: https://komonews.com/news/consumer/charles-viancin-silicone-lids-do-they-work*

Thompson, D. (2013, September 13). *Can Your Language Influence Your Spending, Eating, and Smoking Habits?*. Retrieved from The Atlantic: *www.theatlantic.com/business/archive/2013/09/can-your-language-influence-your-spending-eating-and-smoking-habits/279484*

Thompson, D. (2014, November 14). *Forcing Americans to Save Money*. Retrieved from *The Atlantic: www.theatlantic.com/business/archive/2014/11/save-more-money-everyone/382306*

Titericz, G. (2013, August 13). Data Scientist. (R. Barkman, Interviewer)

Tomislav Stankovski, P. V. (2014, February). Coupling Functions Enable Secure Communications. *Physical Review*, p. 1103.

Trivedi, B. P. (2002, December 9). *Fish Ear Bones Hold Clues to Migration*. Retrieved from *National Geographic Today: https://nationalgeographic.com*

Tuthill, K. (2003, November). John Snow and the Broad Street Pump-On the Trail of an Epidemic. *Cricket*, pp. 23–31.

Tymkiw, C., (2021, May 17). "How COVID-19 Changed Our Saving and Spending Habits," Investopedia. *www.investo-pedia.com/how-covid-19-changed-our-saving-and-spend-ing-habits-5184327*

Tyson, Neil deGrasse. (2015). Cosmos: A Spacetime Odys-sey—transcripts (Episode 3), When Knowledge Conquered Fear.

Underhill, P. (1999). *Why We Buy*. New York: Simon and Schuster.

UNICEF. (n.d.). *Patterns in Conflict: Civilians Are Now the Target*. Retrieved from UNICEF: *www.unicef.org/graca/patterns.htm*

United Parcel Service. (n.d.). *Pressroom*. Retrieved from UPS: *www.pressroom.ups.com/Fact+Sheets/ci.Saving+Fu-el%3A+UPS+Saves+Fuel+and+Reduces+Emissions+the+%22Right%22+Way+by+Avoiding+Left+Turns.print*

University of Kansas. (2016, August 23). *Research verifies a Neandertal's right-handedness, hinting at language capacity*. Retrieved from University of Kansas: *https://news.ku.edu/2012/08/23/research-verifies-neander-tals-right-handedness-hinting-language-capacity*

Valentine, K. (2013, March 12). *Climate Change– As CO2 Emissions Rise, So Will Pollen Counts And Asthma Attacks.* From Climate Progress.

Vanderbilt, T. (2012, September). *How Biomimicry is Inspiring Human Innovation.* Retrieved from *Smithsonian.com: www.smithsonianmag.com/science-nature/how-biomimicry-is-inspiring-human-innovation-17924040/?no-ist*

Vaux, B. (2002). *Which of these terms do you prefer for a sale of unwanted items on your porch, in your yard, etc.?* Retrieved from Dialect Survey: *http://dialect.redlog.net*

Vettel, P. (2015, November 15). *No-tipping policy begins at NYC restaurant and industry is watching.* Retrieved from *The Chicago Tribune: www.chicagotribune.com/dining/ct-no-tipping-nyc-restaurant-danny-meyer-20151119-story.html*

Volkmer, K. (2013, July 17). *What Makes Coca-Cola an Iconic Brand? And How Do I Build One?*

Waldman, A. N. (2012). *Words Can Change Our Brain.* New York: Plume.

Walker, S. C. (2010, March 19). *Elliot Wave International.* Retrieved from Can You Use the Wave Principle to Trade Individual Stocks?: *www2.elliottwave.com/freeupdates/archives/2010/03/19/Can-You-Use-the-Wave-Principle-to-Trade-Individual-Stocks.aspx#axzz3crG4a5RY*

Wansell, G. (2008, April 8). *Revealed: How Autistic Genius Stephen Wiltshire Drew His Amazing Picture of London's Skyline.* Retrieved from *Daily Mail: www.dailymail.co.uk/news/article-557942/Revealed-How-autistic-genius-Ste-*

phen-Wiltshire-drew-amazing-picture-Londons-skyline.
html

Weart, S. (2012, August 17). *The Discovery of Global Warming [Excerpt]*. Retrieved from Scientific American: *www.scientificamerican.com/article.cfm?id=discovery-of-global-warming*

Websdale, E. (2011, July 8). *5 Wonderful Things Inspired by Nature*. From Green Buzz.

Weeks, J. (2009, March 22). *Climate change comes to your backyard*.

Weeks, J. (n.d.). *USDA revises its plant hardiness map, bringing climate change down to earth for millions of households across the country*. From USDA revises its plant hardiness map, bringing climate change down to earth for millions of households across the country.

Weems, S. (2014). *Ha!- The Science of When We Laugh and Why*. New York: Basic Books.

Weinstein, A. (2015). *Through Biased Lenses: The Public Perception of Science Has Changed, but the New View Is No Better*. Retrieved from The American Society for Cell Biology: *www.ascb.org/careers/through-biased-lenses-the-public-perception-of-science-has-changed-but-the-new-view-is-no-better*

Wells, M. (2003, September 1). *In Search of the Buy Button*. *www.forbes.com/forbes/2003/0901/062.html?sh=854ed1b73535*

Werker, J. (1989, January–February). Becoming a Native Speaker. *American Scientist*, pp. 54–59.

Werker, J. (2003). Baby Steps to Learning Language. *Journal of Pediatrics*, S62–S69.

Wessels, T. (1997). *Reading the Forested Landscape*. Woodstock, Vermont: The Countryman Press.

Wessels, T. (2008, January 10). Reading the Landscape patterns. (R. Barkman, Interviewer)

White, T. (2014, March 31). *Strategy. Innovation. Brand*. From Travis White: *http://traviswhitecommunications.com/tag/jokes-as-patterns*

Wikipedia. (2012, October 5). Retrieved from Wikipedia: *https://en.wikipedia.org/wiki/View_of_the_World_from_9th_Avenue*

Wikipedia. (2014, May). *Integrated circuit*. Retrieved from Wikipedia: *https://en.wikipedia.org/wiki/Integrated_circuit*

Wikipedia. (2016, September 16). *Halley's Comet*. Retrieved from Wikipedia: *https://en.wikipedia.org/wiki/Halley%27s_Comet*

Wikipedia. (2016, October 10). *Hybrid electric vehicles in the United States*. Retrieved from Wikipedia: *https://en.wikipedia.org/wiki/Hybrid_electric_vehicles_in_the_United_States*

Wikipedia. (2016c, October 21). *Living Building Challenge*. Retrieved from Wikipedia: *https://en.wikipedia.org/wiki/Living_Building_Challenge*

Wikipedia. (n.d.). *Poliomyelitis*. Retrieved from Wikipedia: *https://en.wikipedia.org/wiki/Poliomyelitis*

WikiQuote. (2016, September 10). *Isaac Newton*. Retrieved from WikiQuote: *https://en.wikiquote.org/wiki/Isaac_Newton*

Williams, A. C. (1992). *The Laboratory Revolution in Medicine*. New York: Cambridge University Press.

Williams, M. (2015, December 23). *What is Halley's Comet?* Retrieved from Universe Today: *www.universetoday.com/48991/halleys-comet*

Willis, J. (2008). *Teaching the Brain to Read*. Alexandria, VA: Association for Supervision and Curriculum Development.

Wiltshire. Stephen (n.d.). Retrieved from Wikipedia: *https://en.wikipedia.org/wiki/Stephen_Wiltshire*

Wiseman, R. (2007). The Search for the Funniest Joke in the World. In R. Wiseman, *Quirkology* (pp. 179–225). New York: Basic Books. From Laugh Lab.

Wolf, A. (2002, November 25). The Olden Rules. *Sports Illustrated*, pp. 118–126.

World of Coca-Cola. (n.d.). *About Us*. Retrieved from World of Coca-Cola: *www.worldofcoca-cola.com/about-us*

Wun, D. (1989, June 12). The influence of culture on handedness. (R. Barkman, Interviewer)

Xue, K. (2014, January–February). *Popular Science*. Retrieved from Harvard Magazine: *http://harvardmagazine.com/2014/01/popular-science*

Yankee. (2007, July). *Lyme Disease Treatment | One Woman's Quest for Answers*. Retrieved from Yankee: *https://newengland.com/yankee-magazine/living/pests/lymecountry*

For additional reading on patterns, visit *www.clearmessagemedia.org/additional-reading-for-see-the-world-through-patterns-by-robert-barkman*.

INDEX

Page locators in *italics* indicate figures.

ABOUT THE AUTHOR

Dr. Bob Barkman retired from Springfield College as a distinguished professor where he taught and did research for 44 years. He served as department chair for both the Science and Education departments in separate years. Dr. Bob was the principal investigator for several projects funded by the National Science Foundation, Massachusetts Department of Education, and others. He took leaves of absences to work as a researcher for the National Marine Laboratory in Rhode Island and later as project director for the Technical Education Research Center (TERC) in Cambridge, Massachusetts. For 10 years, he led an educational program for young adolescents called Real World Science that was nominated to receive the Governor's Spirit of Innovation award in education. His most recent book, Science Through Multiple Intelligences; Patterns That Inspire Inquiry, was recognized by the National Science Teachers Association as "only the best." For his teaching and leadership, Bob was awarded the Sears Roebuck Foundation Teaching Excellence and Campus Leadership award. Bob lives with his wife, Dawn, in Longmeadow, Massachusetts.

Printed in Great Britain
by Amazon

80992162R00188